William Heath Byford

A Treatise on the Chronic Inflammation and Displacements of the

Unimpregnated Uterus

William Heath Byford

A Treatise on the Chronic Inflammation and Displacements of the Unimpregnated Uterus

ISBN/EAN: 9783337407254

Printed in Europe, USA, Canada, Australia, Japan

Cover: Foto ©berggeist007 / pixelio.de

More available books at **www.hansebooks.com**

A TREATISE

ON

THE CHRONIC INFLAMMATION

AND

DISPLACEMENTS

OF THE

UNIMPREGNATED UTERUS.

BY

WM. H. BYFORD, A.M., M.D.,

PROFESSOR OF OBSTETRICS AND THE DISEASES OF WOMEN AND CHILDREN IN THE CHICAGO MEDICAL COLLEGE; AUTHOR OF "THE PRACTICE OF MEDICINE AND SURGERY APPLIED TO THE DISEASES AND ACCIDENTS INCIDENT TO WOMEN," ETC., ETC.

SECOND EDITION, ENLARGED.

WITH NUMEROUS ILLUSTRATIONS.

PHILADELPHIA:
LINDSAY & BLAKISTON.
1871.

Entered according to Act of Congress, in the year 1870,

BY LINDSAY & BLAKISTON,

In the Office of the Librarian of Congress, at Washington, D. C.

SHERMAN & CO., PRINTERS

TO

JOSEPH MADDOX, M.D.,

THIS WORK IS

Dedicated,

AS A GRATEFUL REMEMBRANCE

FOR THE MANY INVALUABLE AND TIMELY ACTS OF KINDNESS

BESTOWED UPON THE

AUTHOR.

1

PREFACE TO SECOND EDITION.

IN preparing the Second Edition of "Chronic Inflammation and Displacements of the Unimpregnated Uterus," it has been the object of the author to add to its usefulness by thoroughly revising and correcting, enlarging and illustrating it. The experience of the profession in the last six years has been sought after, and as faithfully represented as practicable in a work of so limited extent, written under the pressure of many engagements, and much other work. In the lapse of time since the first edition was written, the author's observations have served to confirm the general tenor of the doctrines taught at that time, in reference to the sympathetic influence of the uterus, and the effects of local treatment in the cure of the secondary affections thus arising. And he has been careful to modify the local measures recommended when thought best, and add such additional resources in treatment as have been proven to be useful.

CHICAGO, January, 1871.

CONTENTS.

	PAGE
PREFACE,	5

CHAPTER I.
GENERAL CONSIDERATIONS, . . .   17

CHAPTER II.
SYMPATHETIC ACCOMPANIMENTS OF UTERINE DISEASE,   26

CHAPTER III.
LOCAL SYMPTOMS, ...   47

CHAPTER IV.
ETIOLOGY, ...   62

CHAPTER V.
PROGNOSIS, ...   68

CHAPTER VI.
COMPLICATIONS OF INFLAMMATION OF CERVIX, .   78

CHAPTER VII.
POSITION OF INFLAMMATION, . .   89

CHAPTER VIII.
PROGRESS AND TERMINATIONS, . . .   96

CHAPTER IX.
DIAGNOSIS, ...   101

CHAPTER X.

GENERAL TREATMENT, ... 127

CHAPTER XI.

LOCAL TREATMENT, ... 154

CHAPTER XII.

NITRATE OF SILVER AND ITS SUBSTITUTES, . . 169

CHAPTER XIII.

TREATMENT OF SUBMUCOUS INFLAMMATION, . . 195

CHAPTER XIV.

DISPLACEMENTS, THEIR PHILOSOPHY AND TREATMENT, . 208

APPENDIX.

CASE I.—Case of Ulceration of Os and Cervix, with Inflammation extending into the Cavity of the Neck, attended with Chronic Diarrhœa, ... 229

CASE II.—Ulceration—Severe Constitutional Suffering—Sterility—Complete Cure with Local Treatment alone, . 231

CASE III.—Endocervicitis without Visible Ulceration—Severe Symptoms cured by Local Treatment alone, . . . 232

CASE IV.—Case of Ulceration of the Os and Cervix Uteri, Enlargement, and Retroversion, with some Tenderness, without any Local Symptoms, and with an exaggerated state of Nervous Excitement, ... 235

CASE V.—Ulceration of the Os and Cervix Uteri, with Inflammation extending through the Cavity of the Cervix into the Cavity of the Corpus Uteri, ... 237

CASE VI.—Severe External and Internal Inflammation of the Cervix, that would not bear Nitrate of Silver, cured by the Substitutes for it, ... 239

INDEX, ... 243

DISEASES

OF THE

UNIMPREGNATED UTERUS.

CHAPTER I.

GENERAL CONSIDERATIONS.

There is a large list of symptoms called nervous, or sympathetic, which, although not exclusively confined to women, are more frequently found to manifest themselves in them. They were formerly regarded either as independent affections, or as having various sources of origin; and although hysterical was the term usually applied to them in some of their manifestations, it was not definitely known in what manner they originated. Patient investigation in late years has given us more definite and correct notions of them, and we have come to regard them as nearly always dependent on trouble of some kind in the sexual system. Medical men, however, are not united in their opinion that the symptoms alluded to are thus caused; and they are divided into two well-defined parties with respect to uterine pathology.

1st. There are those who believe that the uterus has very little sympathetic influence in the system; that the diseases of that organ are more frequently the result of diseases in other organs, than of independent origin; that the symptoms accompanying, and almost always found in connection with actual lesion of the uterus, do not at all depend upon this organ; that these symptoms may be cured without any attention to the condition of the uterus, and in fact, whatever cures them, almost always cures the affections of that organ.

2d. The other party holds the opinion that the sexual system of the female, in a state of disease, exercises a very morbid influence over nearly the whole organization. That this morbid influence is particularly exerted over the spinal and cerebral nervous systems; and that the only sure and permanent relief is found in the cure of the disordered condition of the uterus.

Those who adhere to the latter view may be classified under two subdivisions: one of which holds that the sympathetic influence of the uterus is only manifested when that organ is inflamed or ulcerated, and that the cure of the inflammation and ulceration relieves the symptoms. The other maintains that inflammation and ulceration are only of slight, if indeed of any importance; while the cause of all the difficulty is some sort of displacement.

It will probably surprise the student when he is told that all these diverse and various opinions are held by gynecologists of equal eminence, integrity, and opportunity for observation. There is reason for surprise in this consideration, and yet this same diversity of opinion exists in all departments of medicine: for example, as to the nature and treatment of inflammation; as to the essential nature of typhoid fever and its treatment; as to the local or general origin of cancer, and the propriety of extirpation. How can this discrepancy be accounted for? It is not my purpose to answer this question at length, but merely to indicate a few obvious considerations, of which one is, that the attention of medical men has been too recently directed with sufficient intensity to the points involved, to enable it to make an induction full enough to convince by its results all the members of the profession of the correctness of any one view. This, therefore, is just the time when we meet with conservatism in the views of temperate and judicious investigators, as well as with the less laudable conservatism of those who have lived too long to improve. Another consideration is, that while judicious practitioners hold antagonistic opinions as to the nature of diseases, they pursue so nearly the same line of practice as to lead to similar results in the treatment of them. A third consideration relates to the power of prejudice, which forms in very many minds an invincible barrier against the acquisition of truth; and the opinions imbibed in early education are those which are maintained the most persistently, sometimes in consequence of an unwillingness to learn, and sometimes even against the light of reason itself. From the pernicious influences of association and prejudice neither learned nor unlearned are exempt.

My views concur with those who believe in the great sympathetic influence of the uterus, and who consider inflammation and its accompanying effects to be the conditions upon which its sympathetic energies depend. Those who deny to the uterus much sympathetic importance in a state of disease are compelled to acknowledge it under states of increased vital energies. I think it is inconsistent to express a doubt of the sympathetic influence of an organ in a state of disease, while we admit that the same organ, when laboring under unusual vital excitement, causes an exaltation, depression, or depravation of function in many important organs in the system. The stomach, when laboring under the stimulus of digestion, influences to a great degree some of the important organs of the body. The brain is always more or less influenced by digestion; when the stomach is strongly engaged, the brain is overwhelmed, and sleep is unavoidable. This is well exemplified in the torpidity of the serpent: when gorged, he is helpless. When the brain is profoundly engaged, digestion is imperfect and sometimes wholly arrested. In a state of disease there is also a close sympathy between these two organs. When digestion is taking place, the secretions of the kidneys are different from their state at other times: and this is not so much on account of the change in the composition of the blood (for this difference occurs too early to be due to such a cause) as it is owing to the influence of digestion on the innervation of those organs. There is very close sympathy between the kidneys and the stomach in a state of suffering. I cannot remember a case in which two organs sympathizing in their healthy functions do not more obviously affect each other in a diseased condition. Why may we not, therefore, reasonably infer this to be the case with the uterus and other organs? It is interesting to notice some of the physiological and pathological effects evidently caused by the changes going on in the genital system of both male and female. All physiologists agree that without a development of the genital organs, particularly the testes and the ovaria, there is permanent nullity in the characteristics of the individual. When the menses make their appearance they bring with them a long list of physical and functional changes, and at each periodical recurrence there is more or less nervous and functional derangement. Disease results when this process is arrested. Greater effects are produced by pregnancy. A case reported by Dr. Tyler Smith, in the "Transactions of the Obstetrical Society of London, for 1859," exhibits uterine sympathies in a strong light. Dr. Smith says: "In the early part of September (a little over two months from the probable time of conception), after she had been six weeks in the hospital, I was asked to examine her, the probability of pregnancy having suggested itself. She was at this time in a state of extreme emaciation: the vomiting was constant, the pulse ranged from 120 to 140; there was great tenderness of the epigastrium; delirium occasionally supervened. At other times her state was one of semi-consciousness. She lay helplessly in the supine position, unable to move her body and limbs, from profound debility. I ascertained that the catamenia had last appeared about a fortnight before she came to the hospital (18th of July). She confessed after much denial, to intercourse on two occasions shortly after this menstruation. The sickness came on suddenly and had continued without intermission. The uterus was found, on a digital examination, to be enlarged and the os uteri softened. The areola were rose-colored and the follicles somewhat developed, and the mammæ were full and rounded, the development of the breasts contrasting in a remarkable manner with the atrophied condition of the body generally. These facts rendered the existence of pregnancy so extremely probable, that she was subsequently placed under my care and removed to the Boynton ward." On a regimen consisting of one teaspoonful of beef tea alternated with the same quantity of milk the vomiting ceased; a gradual increase of these articles improved the condition of the patient until the 3d of December, when she aborted. She did well for two or three weeks after the abortion, when symptoms of acute phthisis appeared, and she left the hospital in February, 1860, in an advanced stage of consumption. When in her lowest condition, soon after putting her on the beef tea and milk, Dr. Smith describes her as follows: "The pulse continued high, and other symptoms of exhaustion remained without abatement. For many days it was impossible to determine whether she was in that state of pause from vomiting produced by exhaustion, which has sometimes been found to precede death in such cases, or whether the stomach was slowly regaining its tone. Bed-sores appeared on the hips and nates; the process of emaciation continued, and on the 16th of September, her weight was only *forty-seven and a half pounds*. . . . . . . I am not aware of any instance on record of such a light weight. Before the commencement of her illness, she is represented to have been plump and in good condition." Dr. Smith adds: "In some of the worst and most dangerous cases of vomiting from irritation of the gravid uterus, the peril occurs at such an early period that pregnancy may not be suspected, and if the suspicion be entertained, it is difficult to diagnose it with certainty. I believe that many fatal cases occur from this cause in hospital and private practice without their real nature being suspected. There is nothing in the whole range of physiology or pathology more extraordinary than the fact that the gravid uterus, without itself being the seat of special pain, irritation or disease, shall excite fatal disease by reflex irritation in some distant organ. In this way pregnant women may be destroyed by secondary disease of the brain, heart, lungs, kidneys, stomach, or intestines. In fact, there is in particular cases no limit to the poisonous influence exerted on the rest of the economy by the gravid uterus."

He would be an obstinate skeptic in pathology, who, having once observed such a case, or having read the above graphic sketch of it should ignore the uterus in his pathological estimation. Can it be possible that an organ so potent in evil work upon other organs under a state of physiological hyper-excitement, does not exert any bad influence in inflammatory hyper-excitement? I think there is no doubt but that it does do so; and I speak after an observation of a large number of unmistakable cases, that the unimpregnated diseased uterus does produce grave and even fatal disorders in other parts of the organism by its reflex or sympathetic influence, while the organ itself is not suspected to be the original cause of the wide-spread disorder. This is also the testimony of others who have made these diseases a special study.

In well-marked cases of inflammation and ulceration of the uterus there is almost always a long list of accompanying diseases, and they are apt to be of a particular sort. These diseases are generally nervous, but sometimes consist in functional aberrations in some of the important vital organs.

It is curious and instructive to notice the similarity of symptoms excited by a diseased uterus and those arising from spermatorrhœa; and I think that the opinion of the sympathetic connection between inflammations and ulcerations of the uterus and other organs is much strengthened by the fact that, in the male, a slight inflammation or an irritation in the urethra excites much ruinous disorder in the system at large. This urethral inflammation, like uterine, does not often lead to fatal damage. It may be said that very extensive urethritis does not generally produce this effect: even chronic gleet does not produce it. But then chronic gleet as well as other extensive urethritis affects a different part of the urethra. May there not, therefore, be very similar pathological relations between the portions of the genital canal affected in the two instances and the general system in the two sexes? Or will it be said that this kind of urethritis arises from general conditions? I think, considering the well-known and acknowledged causes of spermatorrhœa, such an assertion will not be made. The similarity of the two cases affords an argument in favor of the efficacy of local causes in producing uterine inflammations, and of the powerful and general sympathetic influence of them when once originated.

In order that the similarity may be the more apparent, I subjoin an abstract of some of the most common sympathetic influences of the two, and place them in juxtaposition for convenience of comparison:

Uterine Disease.	*Spermatorrhœa.*
Sterility.	Infecundity, with or without impotence.
Absence of sexual desires.	
Absence of fever.	Absence of fever.
Indigestion.	Indigestion.
Intestinal flatus, cramps, and pains.	Intestinal flatus, cramps, and pains.

Uterine Disease.	*Spermatorrhœa.*
Sometimes emaciation. Sallowness. A healthy appearance preserved in some cases under severe suffering.	Emaciation, with sallowness and leaden color about the eyes. Patients sometimes preserve a perfectly healthy appearance.
Great languor of capillary circulation.	Coldness of hands and feet.
Embarrassment in the respiratory movement.	Respiration often very much embarrassed.
Apprehension of disease of the heart from palpitation and other irregularities.	Palpitation of the heart, and other alarming derangements of its action.
Debility of muscles, and inability to walk.	Great weakness of muscles, sometimes almost paralysis.
Nervous spasms.	Spasms of epileptic character.
Pains in the loins and legs in the course of the nerves.	Pains in the loins and limbs; nervous shocks of pain.
Sight often bad, and other senses embarrassed.	Senses often seriously affected, particularly the eyes.
Vigilance.	Vigilance.
Cephalalgia.	Cephalalgia.
	Congestion of the brain.
Irritability in place of amiability.	Change of character on account of mental disturbance.
Despondency.	Hypochondriasis.
Failure of memory. Weakening of the mind.	Loss of memory; impairment of the intellect.
Mania.	Insanity.
Scarcely any tendency to spontaneous recovery.	Scarcely any tendency to spontane recovery.

The above comparison of symptoms between spermatorrhœa and uterine disease is not intended to be complete, but merely to call attention more pointedly to their similarity. The more we study them, the more apparent is the similarity in the general effects of these diseases upon the system. I have elsewhere intimated that the disease in the two cases is inflammation of a mucous membrane and increase in the secretion of it, to which in the case of spermatorrhœa is added the product of a gland, namely, the testes.

Another and the most important proof of the general influence of these local affections is the subsidence of the general symptoms after the local disease is cured. It is said by those who deny the local origin of nervous symptoms in the female that the general treatment is such as to insure a cure of the local disease in spite of local irritants. Dr. Bennett and all other judicious writers very properly direct the use of general treatment, yet they state that it is not in most cases essential to a cure; that it is merely palliative, or at most, auxiliary. This may be readily verified by anybody who will observe the effects of both kinds of treatment. I am prepared to say that I do not believe it best, in the majority of cases that come under the observation of general practitioners, to resort to any studied general treatment, and that I believe some of the general management resorted to is injurious instead of beneficial. One measure in particular very generally recommended,—confinement to the horizontal position,—I consider as almost invariably fraught with mischief. Seventy-five per cent. of the cases I have treated have had no general treatment, the local being quite sufficient. It will not do, therefore, to say that my cases were cured by the general treatment advised. I think that general treatment is essential in a few instances only, in conjunction with the local. It is at best but auxiliary. That inflammation and its consequences are the cause of these multitudinous ailments is most satisfactorily proven, I repeat, by their entire removal by this local treatment. Experiment is the best means of convincing an impartial practitioner of the accuracy of the theory of the local origin of so many general symptoms. Let him take his local remedies, and apply them diligently and perseveringly, and arrive at his own conclusions; in appropriate cases his success will be the legitimate result. It is to be feared that the direful effects of the white-hot iron, Vienna paste, potassa fusa, &c., much more common in the imagination of the writers than in fact, have deterred a large body of the profession in America and in Europe from making a temperate use of what is called the caustic treatment, and thus have retarded progress in the management of these cases. Nobody will deny that all these agents can, and in injudicious hands do, produce mischief. This is no argument, of course, against their use in proper cases. Those proper cases are the extremely obstinate ones, and such as are incurable by other means. The advocates of the local treatment of uterine disease are not to be held responsible for the rashness of the ignorant. No kind of treatment can be properly advocated or represented but by competent and faithful practitioners; and their conclusions cannot correspond if they are not based upon like trials in similar cases. I cannot resist the conviction, after a careful perusal of what I have seen written against the local origin and treatment of uterine diseases, that the experiments of the writers were not made with sufficient thoroughness; and I think that some of them have allowed themselves to be content with imperfect trials in consequence of preconceived opinions. A judgment is only valuable which has been founded on thorough treatment; and as it requires very considerable experience, and a correct knowledge of the anatomical, physiological, and pathological appearances of the mouth and cervix uteri, I am convinced that errors arise unknowingly by misinterpreting what is seen; that, in other words, we do not always know when the pathological has entirely given place to the physiological and proper anatomical appearances. I have been often asked by medical men, why it is, that after a woman has improved to a certain point under the influence of local treatment, giving promise of a satisfactory cure, all progress ceases, and the cure remains imperfect. In many instances I have had an opportunity of examining the cases which were the cause of these questions. I have found that there was still sufficient inflammation to account for the state of the case, and further local treatment removed the impediment to a cure and perfected it. It would be contrary to all other instances in which general or secondary affections arise from sympathetic influence, if some of the secondary affections did not outlast the primary disease. Accordingly we find that in some cases of long standing uterine disease, the organs affected by it become permanently diseased, and after the cause is removed, require independent treatment for their relief. No proper objection can be urged against the theory in consequence of this fact, as it is only in accordance with other examples, as has been already stated. The cases in which the general symptoms do not subside, however, after the cure of the local (when the former are the consequence of the latter), are not very frequent exceptions to the general rule, that to remove the cause is to cure the disease. And when the general symptoms are not cured, the condition of the patient is generally, if not invariably, improved by the removal of some, and the amelioration of other symptoms.

CHAPTER II.

SYMPATHETIC ACCOMPANIMENTS OF UTERINE DISEASE.

I would not deem it necessary to go into a detail of the particular sympathetic accompaniments of diseases of the uterus, were I not convinced that they are often considered independent affections, and their origin not suspected by very many practitioners. It is my wish to impress the conviction that these diseases are overlooked, misunderstood, and neglected; and that an immense amount of suffering is now borne as a necessity by women, that might be relieved, if we would investigate and study their ailments with as much patience as, and with no more reserve than, we approach and investigate lung diseases or throat affections.

Dr. Scanzoni* says: "The sympathetic phenomena which very distant organs so often present during the course of uterine diseases are of the highest scientific importance." They are the more important, because our attention is more frequently called to them than to their original exciting cause. The secondary or sympathetic diseases often distress patients most, and the fact of their mentioning no other troubles may, without inquiry, mislead us into the opinion that they are independent affections.

We will be able to study the general symptoms of uterine disease more profitably, by taking them up separately as they are manifested by different organs, and without attempting absolute correctness in this respect, it will be practicable to present them in something like the order of frequency in which they occur.

Sympathy of the Stomach.—The stomach is apt to be disturbed as early and as frequently as any other organ by uterine disease. This is no more than we would expect, considering how often and intensely it is influenced by pregnancy, and its great readiness to complication in most affections of other parts of the sys-

---

* Diseases of Females.

tem. Simple anorexia is one of the most common of the sympathies of the stomach, as is also its contrary, voracity; but occasional unbecoming and even disgusting depravity of appetite is also not uncommon. Inappetency sometimes proceeds to the extent of loathing of food and to longing for inappropriate articles of diet. Nausea, with loathing of food and disgust at the smell of it, is another feature of stomach trouble; also frequent vomiting when the stomach is full; an absence of discomfort when it is empty, and the vomiting is sometimes worse when there are no ingesta, and nothing is expelled but some of its secretions, which are usually acid, but sometimes bilious. Gastralgia may occur when the stomach is empty; or during digestion, or immediately after swallowing food. The capacity of the stomach to digest food of any kind is often impaired, but more frequently some particular sort of food disagrees with the stomach and embarrasses digestion; in short, almost every form of disordered stomach may be looked for as the result of the sympathetic influence of diseases of the uterus upon that organ. The grade of functional disturbance may vary from the slightest inconvenience to that complete arrest of digestion which rapidly induces inanition and death. Extreme cases of indigestion, however, are not of frequent occurrence, and the disturbances are rather those of great annoyance than such as result in very serious impairment of nutrition; and many patients who constantly complain of suffering very severely from sensitiveness connected with digestion, attain to a state of apparent robust *embonpoint.*

Sympathetic Disease of the Bowels.—The bowels probably sympathize in diseases of the uterus next in frequency to the stomach, and their functional derangements are multitudinous. Constipation is very common. The bowels, in many instances, have apparently no natural tendency to move. I have one patient, who assures me that she has often been fourteen days without any fecal discharge whatever, and that she dare not try how long she could go without it, but says that she always uses some means to promote the alvine evacuations. In other cases, constipation terminates with diarrhœa; and an alternation of diarrhœa and costiveness, which lasts from two to six days, is a constant and habitual state with the patient. In cases of constipation resulting from this cause, the constipation seems to depend upon a want of muscular tone in the intestines; peristaltic action is deficient, and the appearance of the evacuations is in all respects natural, and there consistence proper. In other cases the secretions are deficient, and the stools are dry, hard, and small in quantity. But constant diarrhœa and irritable bowels are also frequent accompaniments of uterine disease. The passages may be profuse, watery, and exhausting; or profuse and fecal. A peculiar kind of discharge in cases of diarrhœa in uterine disease presents a muco-fibrinous cast of the intestines. The casts are sometimes quite tenacious, and of variable length from two to ten inches, and are often complete casts of the intestinal tube; at other times there are shreds of false membrane of irregular shape and size. The discharge of these substances is usually attended with some dysenteric symptoms. The diarrhœa sometimes seems to be excited or aggravated by certain articles of food; at other times one kind of ingesta seems to agree as well as another; and again the bowels may be quite regular, except at or near the period of menstruation. The irregularity is *often* entirely confined to that time. With or without diarrhœa there may be tumultuous gaseous commotion in the bowels; they may be more or less distended, or without distension there may be annoying borborygmus and motion, from the gas passing from one part of the intestines to another, inducing the opinion that pregnancy exists. The gaseous distension of the abdomen is sometimes so extensive and permanent as to induce the over-willing patient to believe that it is caused by gestation, and being frequently connected with hysterical craftiness, she may impose the same belief on a careless practitioner.

Sympathetic Affection of the Liver.—Closely connected with, and of course, very much influencing the condition of the alimentary canal, is the condition of the liver. Sometimes the bile is poured out in such copious quantities as to induce full and free discharges of it from the stomach by vomiting, and to stimulate the intestines to copious bilious diarrhœa when they are not irritable, but subject to the ordinary stimulation of ingesta. This overflow of bile comes in paroxysms, and produces a sort of cholera morbus. When it occurs only once a month, it is apt to be near the time of menstruation, or it may return several times between the monthly periods. But there is often a persistent absence of secretion for a time, or this condition may alternate with the other; or the bile, instead of finding its way into the alimentary canal, may pass into the circulation and give the skin a jaundiced hue. When the functions of the liver are seriously disturbed, there is apt to be at one time a deficiency of bile, and at another a great redundancy. I have not seen this organ congested to any great extent, as observed by Dr. Bennett. But I have seen an enlargement of the spleen in such instances, though I have not supposed it to be the result of the influence of uterine disease. When copious effusions of bile take place somewhat suddenly, all the pain and spasmodic action observed in bilious colic are likely to present themselves.

Sympathetic Affections of the Nervous System.—Much more distressing, if not more serious suffering, is experienced in the nervous system than in the digestive apparatus. Aches, pains, and complaints of evident nervous ailments, are the peculiar province of uterine disease. There is hardly a disagreeable or even excruciating sensation that these patients do not experience; and too often this real suffering is mistaken by the friends for imaginary, and the patient's complaints are treated with unreasonable impatience and rudeness by persons from whom she ought to receive kindness and sympathy, because her appearance does not correspond with her morbid sensations, as we are apt to observe them in other examples of disease. It is remarkable, too, and a fact that often impeaches them with insincerity in their complaints—when the uninitiated are the judges—that these patients will pass from a state of excruciating suffering and loud complaints, under a little excitement, to one of actual enjoyment and hilarity, or conversely. The transition from the excitement of private company, or a public party, gives way in a few minutes to a doleful condition of suffering and unappeasable complaints. The inconsistency of the complaints and enjoyments, the incapacities and the performances of these patients are almost characteristic—at least in their sudden alternation—and are inexplicable in any other way than by supposing that the pains in the different organs to which they are referred, are more dependent upon the general nervous susceptibility than upon the organic disease of even trivial character. They are strictly neuralgic in their nature, and confined to the nerve-matter, or tissue of the parts. A great number of the disagreeable sensations and pains appear more frequently in particular parts, and hence may be distinctly referred to in this description.

Cephalalgia.—Cephalalgia, in some form, either partial or general, is a very common attendant upon the nervous susceptibility of uterine patients. Often general, the whole head seems to pulsate and thrill with terrible pain, rendering the patient almost frantic with the intolerable aching. In a few hours general cephalalgia subsides, leaving the nervous energies prostrate for a short time, but otherwise the patient is free from all pain. This subsidence would not be complete if the cephalalgia were anything but nervous pain in the head. The general cephalalgia is often, but not necessarily, attended by nausea and vomiting or other stomachic, hepatic, or intestinal disorders; and may be relieved, when that is the case, by emesis or an alterative cathartic. This is what is commonly called sick headache. The most frequent forms of pain in the head, however, are partial, and confined to some particular part; as hemicrania, confined to the whole of one side, or a lancinating pain in the temple, brow, or eye. All these are very common pains in uterine disease; but persistent or frequently recurring pain in the occipital region, or on the summit of the head, is nearly pathognomonic of uterine disease. It is almost invariably the case that a woman has chronic uterine disease if she complain of persistent pain in either of these regions. The occipital pain I have observed in this connection much oftener than the pain on the top of the head. It is ordinarily a dull aching, that completely unnerves the patient and renders her unfit for her duties for days together; it is usually very persistent, in some patients being almost constantly present, but in other cases only occurring once a month, ordinarily at the menstrual period. The pain on the top of the head is described generally as a burning pain; patients complain that they have all the time a hot place on the top of their heads. This pain is probably more constant in duration in patients that have it, than is any other of the pains about the head. I think I have observed that when patients suffer greatly from pain in the head they complain less of suffering which is more directly referable to the uterus, than when any other symptom seems to exceed the others. Indeed, I have met with patients who were martyrs to these excruciating headaches, that did not complain of anything which pointed directly to the uterus as the origin of their sufferings, and yet upon examination that organ was found ulcerated and inflamed; and when these conditions were cured by appropriate treatment, the headache ceased to annoy them. A remarkable instance of this kind occurred to me several years ago. The patient came to town to consult me about what she called neuralgia. The pain was located in the occiput; it lasted one week in every four (her menstrual week), and when very severe she had hysterical convulsions. This took place at almost every recurrence of the headache. She had no backache at any time; her menses were natural in every respect, as far as I could gather from her history, on which I placed the more reliance, from the general intelligence of the patient. She could walk long distances without inconvenience, had no pains in the hips, groins, or legs; in short, she made no complaint from which I could infer the origin of the nervous suffering to be in the uterus, except that the headache was sure to come on at the time of menstruation. Her uterus was ulcerated and inflamed; and after appropriate treatment, was cured, when the sufferings vanished, and she has since enjoyed complete immunity from them. This woman was about thirty years old, and in the midst of her childbearing period, and it might hence be supposed that the uterus would exercise more sympathy than at any other time of life; but, as the following case will show, this is not the fact: Mrs. ———, 49 years of age, had ceased to menstruate three years before I saw her, but was subject to the most excruciating headache every six or seven days, each attack of which would so prostrate her that she would scarcely recover from one before the next would appear. She had some backache and inconvenience in walking, but these symptoms scarcely attracted her attention amid the terrible sufferings caused by her headaches. Six months' treatment addressed to the uterus alone sufficed to remove all this great trouble, and render the woman comfortable and capable of her duties in life. The overwhelming influence of this terrible cephalalgia on the nervous system seems to occupy so completely the capacities of it, that minor pain is unheeded by it, and no cognizance is taken of the sufferings of the less sensitive but inflamed and mischief-making uterus.

Affections of the Spinal Cord.—The spinal cord seems to partake very much of the sensitiveness of the nervous system, probably more so than the brain. Pain in some portion of the spine is almost universally present in uterine disease. The most common parts in which the pain is situated are the sacral and lumbar. Pain is so general in those regions, that it has come to be regarded as necessary in the estimation of very many persons to establish the probable existence of this affection. The pain is fixed and almost constant; but aggravated by anything that excites the uterine vascular system; as standing or walking for a long time, lifting or jumping, or sudden emotions. Fright, anxiety, or anger, as the patient says, "flies to the back," and aggravates the pain. It is especially apt to be worse during the menstrual congestion. Sometimes walking so much increases the pain as to incapacitate the subject of it from that kind of exercise. An expression often made use of to signify sensitiveness of the back, is "weak back." Women will say, I have not exactly pain in my back, but it is so weak that I cannot move on account of it, or can hardly stand, or cannot arise from a stooping posture. The pain may be fixed in any part of the spine. I have a patient, whose backache is at the junction of the dorsal and lumbar region. In connection with these pains there is often tenderness in the same region, so that pressure causes great complaint. The pain is not only increased in the part pressed upon, but it sometimes darts along the nerves around the body.

Sympathetic Pains in the Pelvic Region.—A number of painful localities is generally found about the pelvis; in the inguinal or internal iliac region exceedingly common. Immediately above one of the groins a constant and fixed aching may be found, which is aggravated by all the circumstances that increase the pain in the back. Most generally there is some tenderness or soreness in the part in which there is pain, which is increased by pressure. The pain sometimes extends to the hip and side of the pelvis. It is much more frequent in the left side, but is often confined exclusively to the right, and less frequently it is in both right and left side alike. In more rare instances, the pain is centrally situated behind the symphysis pubis.

Extension of Inflammation to the Bladder and Rectum.—The patient will often say she has pain in the bladder, or pain in the rectum, and believes that these regions are affected. The two last pains, when complained of, are generally very appropriately stated to be in the bladder and rectum, and are indicative, for the most part, of an extension of inflammation to these two organs. When this is the case, pain accompanies or rather is increased by micturition, or immediately after it the pain may occur. The same remarks are applicable to the alvine discharge; during defecation the pain is increased, or then only occurs. These pains are not, strictly speaking, sympathetic; but occur as consequences of the extension of inflammation. The pains indicate correctly the locality of the inflammation. In the iliac region it sometimes extends up the side as far as the mammary region: or there may be pain in this latter place not connected with the former. The pain may likewise be situated between these localities and be independent of any pain in them.

Affections of the Sciatic and Anterior Crural Nerves.—Pain in the course of the sciatic obturator or anterior crural nerves is very common in uterine affections of an inflammatory nature. They are often so severe and aggravated by exertion as to incapacitate the patient from walking. Particular motions cause pain according to the nerve affected. When the sciatic is the seat of pain, sitting down, especially on a hard chair, increases it, so that the patient resorts to cushions for defence against pressure. Pain in the course of one or more of these nerves is often the most distressing circumstance connected with the case, and it is often treated as neuralgia seated in the nerves, while the cause is not even suspected. The pain may occupy the whole length of the nerve, or it may be confined to its upper or lower parts, or to an intermediate portion of variable length. The part of the limb traversed by the nerve may be tender or not; most frequently there is no tenderness. The pain may be fixed, or darting and transitory. It may be constant or paroxysmal; the patient may enjoy immunity for hours and days, or even weeks, or she may be a constant sufferer from them. They are apt as other pains are to be greater during menstrual congestion than at any other time. The pains emanating from the pelvis are not sympathetic, nor are they probably reflex; but they are caused very likely by pressure of the uterus, or they may be produced by an extension of the inflammation to the nerve-sheaths.

Hyperæsthesia.—Akin to pains in various parts is hyperæsthesia without inflammation; great sensitiveness of particular parts. Tenderness of the scalp is often complained of. The whole surface of the head is so tender as to require great care in dressing it, and no pressure can be tolerated without an effort. Of a similar nature is tenderness along the spine. The different spinous processes in some sections of the column cannot be touched without giving the patient great suffering. Pressure upon these tender vertebræ sometimes causes pain to shoot along the spinal nerves, passing out of the intervertebral foramina in the neighborhood. There is occasionally also general tenderness of the abdomen.

Anæsthesia.—Much less frequently there is anæsthesia of some particular parts. The patient complains of a want of the ordinary sensitiveness in them; or there is a feeling of numbness which lasts for some days, and which recurs so often as to obtain the distinction of a symptom of the case.

The muscular through the nervous system, is in many cases very seriously affected. Cramps and spasmodic action are very frequent in particular cases, and they are confined almost constantly to certain limbs. They occur more frequently in the lower than in the upper extremities.

Spasms.—A worse state of things, however, exists when there are general spasms of the limbs and abdominal walls, and hysterical convulsions. They are apparently induced by fatigue, or occur at the time of menstruation. The patient, after complaining of severe pain in the stomach, falls into a state of general convulsions, which lasts from thirty seconds to some hours, and the patient subsequently sinks into a state of quietude, but not of insensibility. These attacks are usually repeated several times and then subside, leaving the patient in the possession of her usual physical condition, which is one of nervous misery.

Accompanying Manifestations of Moral and Intellectual Perverseness.—During the spasmodic action which, in the majority of cases, have to a critical observer the appearance of being partly voluntary, there is apt to be a singular perverseness of moral and intellectual manifestations, which was on a certain occasion very graphically expressed by a clerical friend in speaking of a patient, by saying that she "seemed to be actuated by an evil spirit." In the midst of great suffering they not unfrequently try to bite and otherwise wound those who endeavor to restrain their violent agitation; they attempt to throw the covering from them with the apparent object of exposing their person, or say some very perverse things. At other times they attempt to imitate the symptoms of some grave organic affection. One patient, by heaving up the lower part of the chest spasmodically at rapidly succeeding intervals, induced her friends to think that she was the subject of violent palpitations of the heart, and therefore that she must be the subject of cardiac disease; she also imitated throbbing of the temples by spasmodic contractions of the temporal muscle. When this throbbing of the temples was very violent, I requested her to hold her mouth open so as to relax those fibres, but she looked up and said very wicked things, and became contemptuously calm. A request to hold her breath when the palpitations were violent, induced her to act in the same way, and caused an instantaneous cessation of them. The great peculiarity in these spasms has always seemed to me to be a guarded cunning, a deceitful and perverted consciousness. To a close observer this is always easily detected. By using the foregoing epithets descriptive of the peculiarity of this kind of hysterical phenomena, I do not wish to be understood as saying that deceit, cunning, &c., are indications of freedom from disease on the part of patients who are thus affected. I think this is not usually the case, but that they are the result of the morbid state of the mind and body. The spasmodic action of the muscles is not contemporaneous in the corresponding extremities as in epileptiform hysteria or epilepsy; but is so irregular as to move the body in many different directions, instead of giving to it frequently repeated similar motions.

Syncopal Convulsions.—There is a singular variety of semi-convulsions or syncopal convulsions, which I have noticed in a few cases that I do not remember to have observed in any other connection. They occur very frequently after they have once seized the patient, as often as three or even six or eight times during the twenty-four hours. They take place in the daytime or at night during the sleeping or waking condition, and do not seem to result from any particular excitement at the time. If the patient is sitting and talking, or is engaged in work, she suddenly ceases and slowly sinks down to the floor; she turns her head to one side, almost ceases to breathe, becomes pale and trembles, sometimes very gently, sometimes violently. This state lasts only for a few seconds; she arouses, looks about confusedly, and although she knows she has had a fit, as her friends call it, she does not remember distinctly anything which passed during the time. As these attacks become chronic, they may be attended with very slight convulsive movements, frothing at the mouth, and sequential somnolence; but, ordinarily, this is not the case. If the patient is attacked in the night while asleep, unless some person observes the attack, it will not be known to have occurred, the patient being unconscious of it. There is generally, however, movement enough to awaken anybody who may be in the same bed with the patient. In all cases of this kind, I have noticed great impairment of memory, particularly of recent circumstances. There is not usually any severe pain in the head or spinal centres; there is, in fact, no prominent painful circumstance apparently connected with the case. Patients having such paroxysms are generally worse at or near the time of menstruating; but sometimes they are quite exempt from them at this time, but have them not long after menstrual congestion is over.

Muscular Weakness.—Extreme muscular weakness—I do not mean that which results from general debility, but of some particular set of muscles—is often present as an accompaniment of uterine disease. This is most frequent in the back and lower extremities, not often in the upper extremities. It is probably imperfect innervation of the part, or it may be some affection of the muscles themselves. I have been inclined to look upon it as partial paralysis, resulting from reflex irritation. More or less numbness of the parts exists in connection with the weakness of the muscles.

Circulatory System.—The circulation and its organs are very often deranged to a distressing degree. Palpitation of the heart is often troublesome, and patients are apt to think themselves the subjects of disease of the heart. We are often consulted solely with reference to this symptom, it having absorbed the attention and awakened the apprehension of the sufferer to such a degree that her other inconveniences were forgotten or overlooked. These palpitations are sometimes attended with pain in the region of the heart, which occasionally shoots up to the left shoulder and down the left arm to a greater or less distance, the distress being so great as to amount almost to angina. The palpitation is worse during nervous excitement. It occurs generally in paroxysms. We meet with instances in which it oftener occurs after lying down at night than at any other time. Sometimes it seems to be increased during digestion. The *sensation* of palpitation does not seem to be at all commensurate with the increased excitement of that organ, and *vice versa*. I have observed instances in which the patient complained of violent palpitation, while the pulse and heart, as far as I could judge, were not at all disturbed. In such cases we might say that the sensitiveness of the heart was increased until its ordinary motions were perceived by the patient. Indeed, the pains and increased irritability of the organs supplied with the great sympathetic nerves, seem to result from increased susceptibility or sensitiveness instead of organic changes. There is also sometimes a sensation of throbbing, as though the blood was passing through the arteries in increased quantities, and with increased force in some parts of the system; this occurs mostly about the head, sometimes in the hands and feet, and occasionally inside the head, apparently in the brain; also about the genital organs. Great irregularity of distribution of the blood is often observable, the hands and feet being uncomfortably cold, and continuing in that state for twenty-four hours at a time. In connection with cold extremities, the head is apt to be hot, or warmer than natural; this heat of the head may also be present when the feet and hands are of the common temperature. The heat about the head and face is sometimes almost constantly present in certain patients, and is the source of great annoyance to them. It is apt to be caused by anything that excites the person. The heat is greatest and frequently exclusively located on the top of the head. I do not think that this sensation of heat arises from any other cause as frequently as from uterine disease, and I am sure it is one of the most common symptoms in such disease. There is great heat complained of in the back of the head also, in many instances, and sometimes it extends along the spine, affecting the whole or only sections of it. Burning in the sacrum and loins is very common. Flashes of heat and flushes of color in the face and head, and even in other parts of the body, are very common and annoying occurrences. The power or nervous energy of the heart may be impaired to such an extent as to render the patient liable to faintness on the application of very slight causes—anger, fear, surprise, or even the more tender emotions overcoming the patient very readily.

Respiration.—The respiratory apparatus is not so frequently or so severely affected as some of the rest of the organization; and yet we often meet with some very curious and considerable deviations from the natural condition of its functions. The constriction about the throat, or the feeling as if a ball rose up to the throat and obstructed respiration, and the feeling as if smoke or dust was in the air which the patients breathe, are complaints we hear almost every day. All these sensations, or any one of them, may be aggravated to an agonizing degree, inducing the fear that the paroxysm may be fatal, and causing the patient to suffer for some moments, and sometimes for hours the horrible sensations of impending suffocation. The breathing may be spasmodic from painful and unnatural contractions of the respiratory muscles. There may also be pleurodynic pains during each ordinary effort of respiration. Imperfect respiration, or partial inflation of one lung, or of parts of the lungs, occasionally occurs. The modification of the respiratory murmur arising from this imperfect inflation of one of the lungs I have observed on several occasions, and not without serious apprehension of the result; but in all cases where this was the only modification of physical sounds, the patients have done well, and the inflation improved as the returning nervous energy of the rest of the system was established. The respiration is not often hurried as a constant circumstance, but occurs temporarily as the effect of excitement from mental or moral emotions. In some cases, amid the tumult of nervous excitement during a paroxysm, I have seen the respiratory efforts increased to sixty in a minute;

and, occasionally, these nervous patients constantly have increased frequency of respiration. There are cases in which cough is a very constant symptom; it is a peculiar, nervous cough, as a general thing, and is excited or made worse by anything that renders the patient more nervous. Sometimes it is difficult to distinguish it from the coughs which arise from insidious affections of the lungs. It is possible that the coughs arising from slight lung difficulties may be aggravated by the nervousness consequent upon uterine disease. I once saw a patient affected with a peculiar nervous cough as the effect of uterine disease, which sounded like the barking of a small dog, and the sound was made at every expiration during the waking condition of the patient, except when the mind was intensely occupied. She was an intelligent young married woman, about twenty years of age. While her whole attention was absorbed, she forgot to cough, but as soon as her attention was relaxed, she habitually produced the same sound. This had lasted when I saw her six months or more. When she was embarrassed by a conversation which related to her case, the sounds became much louder and persistent, appearing in perfect synchronism with every respiratory effort. I must further add, that I did not have an opportunity to treat this patient, nor have I heard from her, so that I cannot give her subsequent history; but the rest of the symptoms plainly indicated uterine suffering, and an examination established the fact that she had ulceration and inflammation of the neck of the uterus. She had never borne children or miscarried.

Sympathy of the Excretory Organs.—The excretory organs also sympathize with the uterus, particularly the kidneys. It has been a long time observed that female patients, in a state of nervous excitement, secrete a large quantity of urine, which is usually limpid, almost odorless and insipid. These qualities are most likely dependent upon the amount of water being so much greater proportionately than the salts: these last scarcely seem to be present at all. It is extremely dilute urine. Uterine patients are very prone to large discharges of limpid urine. This kind of alteration in the functions of the kidneys is, doubtless, indirect, and does not occur except in connection with a greatly excited condition of the nervous system as the medium between the kidneys and the uterus. More considerable deviations, however, are apt to take place; the salts are likely to be increased in quantity compared to the amount of water; or one sort of the salts may be greatly over or under the proper proportions in relation to the others. The urine may be decidedly morbid in its composition. It is probable, too, that the deviation is secondary to derangements of the stomach and liver, but, nevertheless, it is often present. The urine may be highly alkaline, or highly acid in reaction, showing the production, to an unusual degree, of salts having such chemical qualities. The presence of the salts in excess, whether of the one kind or the other, is pretty sure to produce painful micturition and other disagreeable sensations, as burning and smarting in the urethra and bladder. There is no doubt, however, that the painful and disagreeable symptoms may arise as the more direct effect of inflammation of the uterus when the urine is correct in composition; hence the examination of the urine will be necessary to determine the cause of the symptoms. But the urine is often secreted in very diminished quantities also in cases of uterine disease; and that, too, without apparent general febrile excitement. Patients frequently complain of this symptom. Whether there is an increase in the excretory functions of the skin at such time, I am unable to say. The skin is probably not very much affected in its excretory capacity as a general thing, but some very curious deviations have been observed.

Mammary Bodies.—More direct are the effects upon the mammary bodies. They are often highly excited by uterine disease; this is no more than would have been expected from the close sympathetic relations between these organs. Congestion is the most common sympathetic condition. The mammæ increase in size, become hot and painful as a general thing, but sometimes there is no change in their sensible or sensitive conditions. The appearances are natural, but the patient complains of a peculiar and painful condition, not unlike the sensations perceived during the suppurative stage of inflammation; but there is neither tenderness, nor swelling, nor heat, nor other deviation, than the unnatural sensation. Sometimes the breasts are really inflamed. The lymphatic glands in the axilla, and from the axilla to the border of the mammæ, in some cases, become affected at the same time; in other instances, however, they do not partake in the sympathies of the mammæ. They also become tender in some cases when the mammæ do not seem to be excited.

Moral and Mental Derangement.—No more constant derangements, perhaps, occur, than are observed in the mental and moral qualities of the patient. The patient loses the complete control which she has been in the habit of exercising over her emotions, and finds herself becoming despondent, fretful, suspicious, and unsteady in her purpose; whimsical, having desires not before experienced, indulging in thoughts and feelings toward her friends which in her former days she did not entertain. She will often call herself a changed woman. If the source of irritation is not discovered and removed, she loses her strength of will entirely; and, instead of her moral feelings being guided by her will under the influence of a sound judgment, she exhibits indecision, and wavers in matters about which she heretofore had no difficulty in making decisions. She finds herself giving way to peevishness to a frightful degree; nobody can please her. In place of her usual satisfaction in the attentions of her friends, she finds fault with their efforts to make her comfortable. Sourness, moroseness, jealousy, carelessness, timidity, and peculiar perverseness change her nature entirely. Sometimes one class of ideas will seize her whole faculties, and she will scarcely think or talk of anything else. She has no patience with anybody who will not listen to her, and believes everybody to be her enemy who cannot sympathize with her in her imaginary troubles. The different phases of mental and moral troubles under which the patient labors are almost innumerable. As will be seen, this state of things closely borders on insanity, and there is no doubt that insanity is often the result of uterine irritation in patients who are hereditarily predisposed to it. I think I have seen cases of insanity that were excited into activity by the great nervous irritation connected with uterine disease. But in place of this steady deviation from her natural mental condition, the patient may generally be sane, and show an abnormal state of mind only when circumstances occur which are likely to excite her, when she loses all control and indulges in excessive anger. Sometimes, in a fit of despondency or melancholy, she contemplates or even attempts suicide. Or, if her sense of wrongs weighs heavily upon her, and no means of redress shows itself, she thinks seriously of fleeing from what she fancies is the cause of them. Still another sort of paroxysm exhibits the acts of a depraved and indecent nature; so disgusting as to shock the witnesses of them, and in her recollection of them to mortify her exceedingly. The common hysterical paroxysm of crying without a sufficient cause, the indulgence in unbecoming and unseemly levity, rapid alternations of despondency and hope, need hardly be mentioned from their familiarity to every observer. When, in reference to such unbecoming exhibitions, patients are kindly remonstrated with, they will, in general, acknowledge the impropriety of them, but will end with saying, "I cannot help it;" which is the unanswerable and doubtless truthful exposition of their mental condition. Neglect of duty in all the relations of life is one of the phases of their mental state. Sometimes a wilful selfishness, caring for nothing but what they fancy, will make them happy, or conduce in some way to their interests, absorbs their whole mind and governs all their actions. At times there is an intelligent appreciation of the impropriety of their actions.

I have dwelt so long on these general symptoms, and have made so much of uterine sympathies, that I am forced to recall an expression made use of in a notice of Professor Hodge's work on "Diseases of Women," that "if all this is true, it is almost a pity that a woman has a womb;" but I conscientiously believe I have fallen very far short of mentioning all the sympathetic evils resulting from chronic diseases of the uterus, and I only design this as an outline view of a subject that will fill itself up in painfully warm colors in the observation of those who devote themselves to a close study of the diseases of women. While this is my conviction, I do not wish to be understood as saying that nearly all of the above symptoms will show themselves even in a majority of cases; some of them will be prominent in some cases, others in other cases; and in rare instances we meet with nearly all of them in some sufferer, and in nearly all chronic cases we shall find enough to move us to commiseration for the ruined health of women thus affected. I know there are thousands of my peers in the profession, who do not see in the foregoing array of symptoms any indication of disease of the uterus; and when uterine diseases are obviously coexistent, they are not arranged in the order of sequency. This does not shake my faith in the facts I have observed for myself, nor disturb my judgment formed from an observation of a very large number of cases carefully watched through all stages of progress to their termination. That all the above symptoms may occasionally be present in cases in which the uterus is healthy, I have often observed; but that they are also present as the proximate and remote effects of uterine disease, I am well satisfied. Another well-established fact, according to my judgment, is, that the direct symptoms referable to the uterus may be feebly pronounced, while some or even a large number of the sympathetic disturbances are very prominent; and judging from the freedom of pain from other inconveniences experienced in the uterine region, there are even cases in which the uterus does not seem to suffer at all. These cases are well calculated to mislead us, and to induce the opinion that the womb difficulty is of minor importance, and need not be the object of solicitude until we get rid of the more troublesome and prominent symptoms. We cannot be too careful in our consideration and management of this class of cases, and I insist, that while we adopt judicious remedial means for the removal of the more afflicting symptoms, that we must address ourselves to the disease of the uterus, however slight it may appear to be. I have seen too much good result from the observance of this direction not to dwell with emphasis upon its importance. The cure of the uterine disease will be a valuable diagnostic measure in such cases. Not only may there be a great difference or want of correspondence in the severity of the local and general symptoms, but in many cases in which the general symptoms have almost made a wreck of the health and happiness of the patient, the local inflammation and ulceration will be found upon examination to be trifling in amount and degree. The inflammation may be very slight and the patient suffer very greatly from it, either generally or locally, or both; or the ulceration may be extensive and the inflammation very considerable, and yet the patient hardly be sensible of any inconvenience whatever from its presence. This statement will be confirmed by careful observers in this field of research. This, however, will prove a stumbling-block to those who entertain the opinion that uterine disease is of small importance in the consideration of woman's ailments. They seem to think that there is of necessity an exact and invariable seeming correspondence between the magnitude of cause and effect, and they point to these cases and say, the symptoms were present, but a very trifling, if any uterine disease showed itself upon examination; or, they will say there was great ulceration, but the patient did not suffer from its presence, at least not in proportion to the amount of local disease. I need not particularize instances in which other diseases are comparatively latent, or cases in which the symptoms are unduly severe compared to the amount of actual disease, as they will suggest themselves to every intelligent practitioner. But recurring to the sympathies of the uterus, we find that while some patients are not affected at all by pregnancy, and others favorably affected, their health being better then than at any other time, that some absolutely perish on account of the functional derangements inaugurated by pregnancy; and, as is shown on a former page, organic diseases are not unfrequently lighted up. We will probably always be at a loss to understand precisely this difference; but there can be no doubt that it is more on account of constitutional differences than local ones. The concatenation of sympathetic influences may be caused by the greater susceptibility of the organs secondarily affected. In fact, the only mode of accounting for it is by supposing this increased susceptibility. I am convinced that this great but inexplicable diversity of sympathetic effects is as likely to result from uterine disease as from pregnancy. We must, therefore, expect a very great range of difference in the extent of sympathetic derangement from uterine disease. It is interesting to observe the rise and development of the sequences to diseases of the uterus. How far can the uterus produce a direct effect in creating this large amount of sympathetic disorder? Are most of the symptoms produced by the direct sympathetic relation of the uterus to other organs, or does the diseased uterus first affect some other more influential organ detrimentally, and then this last the organism generally? I am inclined to think, from a large observation, that the uterus has close sympathy with only a few organs, and no one probably is so powerfully affected by it as the stomach. It is the first organ affected in pregnancy, being brought into a morbid condition in a very few weeks. The well-known, powerful, and almost universal sympathetic influence exerted by the stomach upon other viscera is sufficient, when it is diseased, to account for the great variety of subsequent symptoms. The stomach is the great centre from which radiate abdominal, thoracic, cerebral, and spinal disturbances almost *ad infinitum*. And there can be no reasonable doubt that it is an active agent in originating the disturbances of the great vital organs. The subject of the sympathetic influence of the uterus then becomes the more interesting and important, from the fact that very slight deviations from its ordinary condition arouses the most influential of all the organs to a state of disease, which depresses the functional energies and increases the susceptibilities of almost all the rest of the organism. In addition to the chain of sympathetic susceptibilities produced by this state of the stomach, too frequently the digestive powers of that organ are impaired or perverted, so as to supply the chyme in deficient quantities or in deteriorated quality, and in this way injuriously affect the composition of the blood, inducing anæmia or oligæmia. Imperfect nutrition will follow, as a matter of course, in the one case, and perverted nutrition in the other, so that emaciation or obesity will be ordinarily present. Another organ, probably, in direct sympathy with the uterus is the cerebellum, as it seems to me to be as frequently affected as the stomach. The mammæ are, of course, in direct sympathetic relation with the uterus, and yet they are not uniformly affected in all cases when the uterus is very seriously diseased. I do not believe that we are able to say at present, whether there are other organs that come directly under uterine influence. A proof of the powerful and very ready effect upon other organs, of irritation of the uterus, may be found in the fact, that very often when the patient is in a condition of comfort, so far as her general suffering is concerned, an application of nitrate of silver to a morbid os uteri will give her excruciating pain in the head, render her exceedingly despondent and irritable, and very much aggravate the symptoms with which she is affected. This I have so often observed to be the case, that I cannot but regard it as one of our diagnostic means. After such an application, the patient will generally complain of an aggravation of the general symptoms, whatever they may have been, and say that all the pains are made worse by the application of the caustic. When an organ has been the subject of irritation or functional derangement for a long time, in consequence of sympathy with the uterus, it may become the subject of organic disease, and then continue as an independent affection of perhaps a dangerous character; or if organic has not succeeded to functional disease, the power of habit, which is so frequently thus engendered, will perpetuate morbid action for an indefinite period after the cause of it has been removed.

CHAPTER III.

LOCAL SYMPTOMS.

The symptoms which more directly indicate inflammation and ulceration of the cervix uteri, should be dwelt upon with some minuteness. Some of the symptoms have been already mentioned, and the rest are reserved for a separate notice.

Pain in the Sacral or Lumbar Region.—Pain in the sacral or lumbar region is one of the most constant, and when persistent, indicates, with a good deal of certainty, disease of some kind in the pelvis. The pain in these regions, which is caused by the uterus, is ordinarily central, being in the middle of the sacrum at its lower extremity. It is sometimes at its upper extremity, or it extends the whole length of the bone. Not unfrequently a painful spot may be found on one side, over the sacro-iliac junction. Some patients describe the pain as if a bundle of nerves were pulled upon from the inside of the sacrum, and others describe it as an aching or burning pain. Accompanying the pain in the sacrum is often a sense of soreness upon pressure, an inability to sit with comfort, on account of the tenderness of the lower part of the sacrum.

Pain in the Loins.—Pain in the loins is probably not so common as that in the sacrum, but is quite as various in its nature. Very frequently there is great weakness in the loins, so great in degree sometimes as to prevent the continuance of the erect posture for any length of time. I have had a number of patients who were unable to stand upon their feet long enough to dress their hair, on account of a weak back.

It is remarkable that patients often feel this weak back more when standing than walking; and they are sometimes able to walk a distance without any great inconvenience, but as soon as they stop, the weakness is apparent to a distressing degree.

Inability to Walk.—Ordinarily, the weakness disables the patient from walking. The pain in the back is almost always increased by walking or standing, and on this account the patients avoid being on their feet, although the back is strong enough. But there are many patients who have severe inflammation of the cervix, who do not experience any of the inconveniences in the sacrum and loins already described; but some of them are very generally present.

Great pain in the back, closely resembling that arising from diseased uterus, is also caused by hemorrhoids, prolapse, or inflammation of the rectum. The pain caused by diseases of the rectum, I think is much more frequent on the left side of the sacrum, and in the left nates or hip, than it is centrally situated; in fact, I have come to regard pain, confined to the left nates and hip, as indicating, with considerable probability, rectal disease, and I always inquire into the functions of that organ when such pain is present. It differs in position from the pain in the iliac region, so common as the result of uterine disease. It is situated near the sacrum, and more in the side of the pelvis than the latter.

Pain in the Iliac Region.—Pain in the iliac region is very common. In frequency, it is next to pain in the back. The pain is commonly situated a little to the anterior superior spinous process of the ilium, and below the level of it. It is not referred to the iliac bone, or fossa, but to a place a little above the groin. We often meet with it on both sides, but much more frequently on one only; on the left side much oftener than on the right. Dr. Dewees considered pain in the left groin, or a little above it, as almost diagnostic of prolapse of the uterus. It is certainly very frequently indicative of inflammation of the uterine cervix.

Soreness in the Iliac Region.—This pain is generally accompanied with soreness upon pressure, and sometimes there is soreness upon pressure when there is no constant pain. Walking, standing, or riding generally increases it. A severe shock or strain, from lifting, will sometimes cause pain suddenly to appear in this region, when it had not before been observed.

Pain in the Side, above the Ilium.—Instead of the pain situated as here described, there is often pain higher up in the side, or in the iliac fossa, or along the crest of the ilium, and even midway between the crest and ribs of the side. These pains are not in the ovaria, although they seem to point to the ovaria more directly than to the uterus, and are by some regarded as a symptom arising from ovarian inflammation. Dr. Bennett admits that it may be a sympathetic painful condition of the ovary. It is not material whether this is true or not; it is certain that it is very frequently present in uterine disease, and almost invariably cured by remedies addressed to the uterus, instead of to the ovaria.

Weight, or Bearing-down Pain, or Uterine Tenesmus.—Another indication of uterine disease, of less frequent occurrence, is a sense of weight in the loins or pelvis. This sense of weight is experienced in the loins and iliac regions more frequently than elsewhere; but it is often felt at the pelvis, and oftener in the perineal and anal regions. Patients express themselves as feeling a heavy weight dragging upon the back and hips, and others feel as though the insides were dropping through the vagina. Occasionally we meet with such urgent uterine tenesmus that the patient is obliged to keep the recumbent posture, in order to enjoy any comfort. In such cases, the patient in the erect position cannot resist a constant desire to "bear down," resembling the tenesmus of dysentery. This sensation is sometimes more distressing than any other symptom, and obliges the patient to desist from walking.

Leucorrhœa.—Leucorrhœa is one of the symptoms usually relied upon as an evidence of disease of the uterus. In the healthy condition of the uterus and vagina, there ought to be no discharge; the vaginal canal is merely moist, and no mucus should make its appearance externally. When the mucous membrane is temporarily excited, there is more than ordinary secretion; but it ceases as soon as the cause of excitement passes.

We should *a priori* expect increased vaginal discharge to be accompanied with some form of disease, especially when it continues for more than a few days. Our knowledge of the discharge from mucous membranes lining the cavities elsewhere, will afford us enough data to confirm these views. We do not expect to see a constant flow, however moderate it may be, from the male urethra, when it is perfectly healthy; and we take gleet as an evidence of chronic urethritis, and it is generally the sequence of an acute attack of that disease. A constant discharge from the nose is an evidence also of more or less disease. It is just so with the vagina. The indications from leucorrhœa are derived from the color or consistence of the discharge, or both. The discharge from the vagina, resulting from mere excitement of the vaginal crypts, is thin, glairy, and not very tenacious. It is ordinarily acid in reaction. There is no color, and but little consistence to it. When a moderate excitement of the internal mucous membrane of the neck of the uterus produces a discharge of mucus, sufficient to appear at the orifice of the vagina, the discharge is white, not unlike milk, and when examined closely, will be found to consist of minute coagula swimming in a little clear fluid. When the mucus flows from the mouth of the uterus, it is thick, and resembles very closely the albumen of an egg, and is alkaline in reaction. When it passes into the vaginal canal, it meets with the acidity of the vagina, and is coagulated, and the whole changed from a colorless translucency to an opaque white. The reason that the coagula are small and so numerous, may probably be found in the fact that the mucus arrives in the vagina in such small quantities; each coagulum represents a minute drop of mucus, changed in quality. As, however, the mucous membrane of the vagina furnishes only a small quantity of acidity, when this alkaline discharge from the cervix is copious it soon neutralizes the vaginal acid, and passing through this cavity unchanged, appears at the external parts possessing its characteristic qualities. We then hear the patient complain of a tenacious albuminous leucorrhœa; she will nearly always compare it to the white of an egg, but state that it is more tenacious. Unless the quantity is considerable, the mucus from the internal cervical membrane does not appear at the external orifice unchanged, but passes into this curdled condition. There is often a considerable quantity of this cream-like leucorrhœa in the whole length of the vagina, and hence it has been supposed by many that this is the vaginal mucus, in its natural condition, and they have called it vaginal leucorrhœa.

Character of the Mucus in the Vagina.—I am satisfied that it is changed cervical mucus; the vaginal fluid does not undergo any kind of coagulation, but appears at the vaginal orifice as a clear, thin, almost watery fluid, which moistens but does not stain the linen. This colorless mucous secretion is indicative of increased vascular and glandular excitement, without detriment to the integrity of the membrane whence it is derived, and the excitement may be due merely to temporary congestion, in which case it will disappear, or it may be inflammatory, when it will become more persistent and possibly permanent.

Amount of Leucorrhœa not always proportioned to Extent of Disease.—The abundance of this discharge is no criterion by which to judge of the amount of inflammation, or its intensity, but it will scarcely remain colorless after the integrity of the membrane is invaded. When the albuminous fluid appears at the orifice of the vagina, there is persistent cervical disease, almost of a certainty.

Yellow Leucorrhœa, when there is Inflammatory Ulceration.—The thick, white, or egg-like albumen will be mixed, when there is ulceration in the cervix, to a greater or less extent, with pus, so that it will be stained yellow if the quantity of ulceration is considerable, and its surface is producing pus, the yellow will preponderate in the color, and sometimes the whole of the production becomes yellow. The yellow color may be in streaks through it, or intimately mixed with it, so as to stain it uniformly; or the pus may be mixed with the white creamy secretion found in the vagina. Pus may be mixed with any of the varieties of leucorrhœa, and impart to it its tint, more or less completely.

Yellow color always a sign of Ulceration.—I am in the habit of considering the pus-colored leucorrhœa as indicative, with great certainty, of destructive or ulcerative inflammation of the mucous membrane of the genital canal. This inflammation may be situated in the vagina, in the cervix of the uterus, or in the uterine cavity; and I can hardly conceive of the production of pus by a mucous membrane with a whole epithelium.

Ulceration sometimes exists without Leucorrhœa.—While I am almost confident of the existence of ulcerative inflammation somewhere, when this purulent leucorrhœa shows itself, and persists for some time, I am not, on the other hand, at all confident that ulceration does not exist, if the yellow leucorrhœa is not present. Indeed, I do not regard leucorrhœa necessary to establish the existence of ulcerative inflammation. There are many cases in which it is quite evident that mischievous inflammation is plainly the cause of the invalid condition of the patient, and yet neither mucus nor pus ever shows itself at the vaginal orifice. This may probably be attributed to two circumstances: first, all ulcerated surfaces do not discharge pus, or if pus is discharged, it may be in very small quantities; in the second place, the absorbents of the vagina may be so active as to take it up before it arrives at the external parts. Cases have been observed in which a large secretion is caused by a small amount of ulcerative disease. Notwithstanding the fact that in uterine disease leucorrhœa is a common and significant symptom, we cannot base an absolute opinion on its absence, in any given case.

How is the Pain produced?—How are the local painful symptoms produced? Is the pain in the groin or ilium caused by prolapsus and traction on the broad or round ligaments? I think not. Pain and sensitiveness in the ilium are so frequently present, when I cannot detect any kind of displacement, and so generally disappear when the inflammation is cured, that I am convinced displacement is not necessary for their production. They are of that character of pains which range themselves in the category of the vague, yet indispensable term, sympathetic; or, of the not less fashionable, yet equally indefinite term, reflex.

Bearing down not caused by Displacements.—The sense of weight or bearing down in the pelvis is one about which there would, from its nature, seem to be no doubt as to its origin being in displacement. It gives the patient the idea that the womb is bearing with unusual weight on unusual places, viz., the perinæum, the rectum, or the bladder; and yet, in a great many instances, we will fail, I think, to detect any deviation from the natural position of that organ, and as soon as the inflammation is cured the symptom vanishes, without any treatment being directed with reference to displacement. How can we account for this symptom? I think its explanation may be found in the fact that the pelvic organs, on account of the inflamed condition of the uterus, and the general pelvic vascular turgescence, are unusually sensitive, and receive painful impressions from contact, which, in the absence of these conditions, would have no effect in causing inconvenience of any kind. I also think that moderate prolapse, retroversion, or other displacement, when unattended by congestion or inflammation, may exist for a long time, without giving rise to any disagreeable sensation whatever. When the uterus is slightly displaced, with considerable pain and sense of weight accompanying this condition, the displacement is commonly considered to be the cause of the distress. When, however, the uterus occupies a normal position, and a sense of weight and pain still exists, they are regarded by most practitioners as the results of an "irritable uterus." That the uterus is sensitive, "irritable," if the term suits better, there is no doubt; but that it is ever so, without congestion or inflammation, I do not believe.

Severity of Suffering not commensurate with Amount of Disease.— The great error in the estimate of the importance of uterine inflammation, is in endeavoring to measure the amount of inflammation by the severity of suffering; in assuming, that because the woman suffers a great deal there must necessarily be extensive inflammation or ulceration. I believe I have seen more nervous prostration, more keen suffering, and have heard louder complaints from a small amount of endocervicitis, than from extensive and obvious external ulceration. Pelvic congestion and increased sensitiveness of the viscera contained in the pelvic cavity, caused by a small amount of persistent inflammation in the neck of the uterus, calls into action, in an exaggerated and intensified form, all the sympathies which are excited by the uterus, in its physiologically congested condition, and its persistence wears the more upon the general organism, on account of the increased sensitiveness produced from day to day, by virtue of its chronicity alone. It is anticipating what I shall say in the chapter on Prognosis, to state that endocervicitis is not only more difficult to cure, but more destructive to the health and happiness of the patient than inflammation and ulceration external to the os. Indeed, we often find cases of extensive ulceration very apparent through the speculum, and consequently entirely unmistakable to the most careless observer, which produces less inconvenience than an amount of endocervicitis so small as to escape the attention of any but an experienced gynecologist. This fact is perplexing, but the knowledge of it will cause a proper appreciation of what is apparently a trifling matter.

Effects on the Functions of the Uterus.—Having given the foregoing sketch of the general and local symptoms of ulceration and inflammation of the neck of the uterus, I purpose to glance at the effects produced on the functional action of that organ. The first function assumed by the uterus and the last it continues is menstruation. It becomes a matter of interest to the physician to ascertain the cause of deviations in a function so persistent, so general, and so important to the health of woman. As inflammation is the cause of injurious and even destructive tissual changes, and of functional aberrations in the vital organs much more frequently than any other pathological condition, so I think that the functional aberrations of the uterus depend particularly much more frequently upon inflammation than on any other cause.

Pain during Menstruation.—Pain during menstruation is not necessarily attended by deviation from the normal monthly flow. That there are varieties of dysmenorrhœa or painful menstruation, with unusual quantities and extraordinary kinds of discharge, is true; but, in many instances, the discharge, though accompanied with pain, is right as to its character and quantity.

Kind of Pain attendant upon Uterine Inflammation.—The kind of pain attendant upon uterine inflammation is, for the most part, the same in quality, but varying in intensity. It is a continuous sore pain, with heat in the parts, sometimes so slight as to give the patient very little inconvenience, and it varies from this to pain of great severity. The pain is at times sufficient to cause the patient to keep her bed for several days, and sometimes for the whole period of the menstrual flow; occasionally it amounts to agony, prostrating her by a paroxysm which may last for hours, or even several days.

Cramping Pain.—Instead of this continuous sort of pain of varying intensity and duration, there are less frequently painful throes "coming and going," like labor-pains or after-pains. This kind of pain is often mistaken for colic. They are often very severe, and may last a few hours, or several days. They may depend on some substance contained in the uterus, as shreds or membranes of fibrous exudation, and cease at their expulsion. But more often no such cause can be discovered in the evacuations; nothing can be found but fluid blood, or coagula evidently formed in the vagina. In other cases, the os internum uteri is small, and does not readily admit the passage of the uterine sound.

Effects of Partial Closure of the Os Uteri on Menstruation.—Many practitioners believe that this condition of the os internum, by preventing the ready flow of the blood, causes it to accumulate until the quantity is sufficient to arouse expulsive efforts for its extrusion. In a large minority of cases, I have had the opportunity of observing there was no coarctation; and in several of the worst cases I have met with, the os internum allowed the sound to pass with so much freedom that I could not distinguish its locality. It is also true that many cases in which the os externum was not larger than a small pinhole, the patients menstruated without any pain whatever. I do not assert that painful menstruation does not occur as the effect of any other cause than inflammation, though my conviction is, that inflammation is its most frequent cause. By far, the most frequent causes of dysmenorrhœa from obstruction I meet with, are in connection with flexions of the uterus. I can easily understand that a sharp curvature in the cervix, or at the junction, will prevent the free efflux of the menstrual fluid. In such cases the pains resemble labor-pains, and are doubtless of the character of uterine contractions. The pain from inflammation may occur at any time during the menstrual flow, and before and after it. Not unfrequently a paroxysm of severe pain, lasting several hours or a day, warns the patient of the approach of the discharge; and subsides suddenly and completely, or gradually and incompletely, as soon as the discharge is fairly established. Frequently the pain continues during the whole time of menstruation, beginning shortly before or synchronous with the discharge, and subsiding with it, though in occasional cases it continues after it. We sometimes meet with patients who begin to menstruate without any suffering, but who have pain during the flow, or after its discontinuance. I think that a majority of patients affected with uterine disease have some pain during menstruation; but there are some who have none whatever, and pass through their period with little or no suffering.

Manner of the Flow modified by Inflammation.—The manner of the flow is often modified. Instead of the continuous flow commencing moderately, gradually increasing, and then as gradually declining, every manner of deviation, almost, may exist. With some, the discharge begins naturally, increases very rapidly, until, at the end of twenty-four or thirty-six hours, an average amount is lost; and then the discharge suddenly declines and ceases, or continues in very moderate quantity for a time longer, and gradually or suddenly stops. With others, the flow may begin and proceed naturally for a day or two, cease for one or two days, and then reappear and flow freely for a sufficient time. When menstruation proceeds in this way it is generally attended with pain. These two varieties are more frequent than any other.

Duration of the Flow.—The duration of the flow may not be affected by it. The flow may continue three weeks or the whole month. This, however, is not frequent. It does not much affect the periodicity of return, of menstrual congestion, and of effort; but it is not unusually the case that we cannot distinguish the discharge which attends ovulation from the hemorrhage which proceeds from an ulcerated surface, as hemorrhagic congestion is so constantly present. We often meet with patients who are so confused by the frequent irregular returns of uterine hemorrhage, that they lose all reckoning as to the time for the menstrual return. Occasionally, continuous hemorrhage is present. The most frequent deviation from regularity in periodicity of the menstruation consists in a slight anticipation of the time of its return.

Menorrhagia.—Menorrhagia or hemorrhage at the menstrual period is not an unusual functional deviation. The hemorrhage is often very considerable, and continues after the usual period has passed by. The flooding is usually greater while the patient is in an erect posture, and it is greatly moderated by recumbency. Occasionally, however, it is not moderated by this means. It would seem probable, *a priori*, that menorrhagia would be the rule with patients affected with uterine inflammation, but such is not the case. I am not sure that even a majority of patients have it.

Menorrhagia frequent in Endocervicitis.—I have observed that menorrhagia occurs much more frequently in patients in whom the inflammation occupies the cavity of the neck; this also is the case with painful menstruation. All cases in which there has been either great pain or hemorrhage, or both, for they are frequently coexistent, have been, in my observation, cases in which endocervicitis is the principal disease. Menorrhagia is not always the result of inflammation of the uterus, though inflammation is its most frequent cause; and in such cases it cannot be cured without first curing the inflammation.

Amenorrhœa sometimes results.—Amenorrhœa is the least frequent of menstrual deviations as the effect of inflammation of the uterus; but this inflammation is frequently the cause of scanty menstruation. It is curious to notice the manner in which this scantiness occurs. It seems to come on after the inflammation has lasted for a considerable time, and is almost always associated with sterility. In cases I have watched for some time, I have been induced to believe that the organ was atrophied and rendered less vascular and erectile; probably on account of a deposition of fibrin throughout the general structures of the uterus. I have not been able to verify my opinion in any case by dissection. The scantiness is sometimes attended with irregularity, which consists in postponement or lengthened intervals. I treated one patient for endocervico-metritis, in whom the uterus did not appear to be, as far as I could measure it per vaginam, more than one inch and a half in length, and correspondingly small in the other dimensions. This patient would menstruate sometimes only a day every month, and discharge but half an ounce of blood each time, and occasionally the discharge would not return for five, six, and even nine months. In early life her menses had been regular in quantity, quality, and times, and unattended with pain. She was barren, having never conceived, as far as she was aware. She dated the beginning of her disease from vaginitis during an attack of fever, which occurred two or three months after marriage.

Function of Generation affected by it.—The great function for which the uterus was formed, that of generation, seems very frequently to be disturbed by inflammation of the neck of the uterus. Some practitioners think, because a woman bears children with frequency, the uterus cannot be much diseased. This is unquestionably a mistake. I have known many women with extensive ulceration bear children very frequently. Conception may be entirely prevented by inflammation, or gestation may be arrested by miscarriage, or labor may be rendered difficult by it. It has already been stated that many women will bear children, having at the same time very considerable disease of the uterus, but there is always great liability to embarrassment of the function in such cases. There is no doubt that many cases of sterility depend wholly upon inflammatory action about the neck.

Sterility.—Sterility is attended by different circumstances. Some women are sterile their whole lifetime; others, after having borne children to the full period and given birth to them, become sterile for years, or for the whole of their subsequent life; others again become pregnant soon after marriage, miscarry at an early period, and never again conceive. In most cases of sterility which I have had the opportunity of examining, I have invariably found evidence of inflammation in the cervical cavity. Very often the inflammation is confined to this cavity. The history of these cases showed that ulceration and inflammation had existed from the time of menstruation; these were cases in which conception had never taken place. In cases of sterility in which the women have become sterile after having once borne children, ulceration is usually situated around the os, extending upward into the cavity of the neck. This is almost certain to be the case if the woman has borne several children. When the patient has miscarried but once, there is not likely to be external inflammation to any great extent; but if there have been several abortions, the ulceration is apt to creep out and manifest itself upon the labia uteri, and sometimes becomes very extensive. Although the foregoing statements with reference to the position and extent of ulceration in sterility will generally be found to correspond with the appearances, yet we must not be surprised to find pretty extensive ulceration external to the os uteri in the originally sterile patient; and in those who have borne children and become sterile afterward, we shall sometimes find no external ulceration. The result of my observation is, that when sterility originates in uterine inflammation, it is in that form of it known as endocervicitis. Sterility is oftener associated with the condition and quality of the leucorrhœal production than on any apparent incapacity of the uterus. In many of these cases the secretions from the vagina are very abundant and intensely acid, so as to produce irritation of the external organs. Although the semen is diluted and defended from the influence of acid vaginal secretions by mucus of alkaline reaction, yet when these vaginal secretions are abundant and possess strong chemical qualities, they may destroy the vitalizing influence of the seminal fluid, and thus prevent fructification. Or the very thick, tenacious, albuminous fluid, which sometimes plugs up the os uteri and whole cervical cavity, may prevent the ingress of the spermatozoa, which, by their independent motion, according to present belief, penetrate the uterus, meet the ovum somewhere on its passage to the os uteri, and produce their fructifying influence upon it; and thus is precluded the possibility of effective insemination.

Abortion.—But conception may readily occur and pregnancy be complete, and after gestation has continued for a certain time abortion may take place. Abortion is a very frequent effect of inflammation and ulceration of the os and cervix uteri. The seat of inflammation or ulceration which most frequently induces it is inside the cervical cavity. We find some patients who have aborted very frequently and never had a full-term child; others, who have had one or more children, but who miscarry every pregnancy afterward; and again, others who miscarry frequently and occasionally go to full term. It is not strange that miscarriages should result from this cause; *a priori*, miscarriage might be regarded as its necessary effect. Nevertheless many patients bear children at term who labor under severe ulceration, and who are prostrated by the constitutional sympathies accompanying pregnancy.

Conditions of the Uterus in Abortion.—Two general conditions of the uterus exist as the effect of the cervical inflammation, and are probably the proximate causes of abortion, viz., congestion or arterial injection of sufficient strength to cause hemorrhage; and, perhaps, by means of insinuation of the clots, separation of the placenta, or irritability of such a nature that contraction and expulsion follow conception; or, perhaps increased sensitiveness of the mucous membrane may increase its excito-reflex influence so as to arouse uterine contraction, and thus cause the fœtus and membranes to be expelled. When abortion is caused by congestion, it is apt to be ushered in by hemorrhage. The hemorrhage, after continuing for a varied length of time, from a few hours to several days, is followed by uterine contractions. When abortion is the result of increased irritability, the first symptom is contraction, with the paroxysmal pains attendant upon it. This continues for a time, when hemorrhage and expulsion succeed. When abortion occurs once, it is very likely to recur in every subsequent pregnancy about the same time until the disease is cured upon which the abortion depends. While abortion is very apt to recur in the congestive or hemorrhagic variety, it is generally not so exact in the time of recurrence. This variety, however, takes place more frequently at the time when the monthly congestion is present, while the other is independent of such influence. The probability is, that in the congestive variety the fœtus perishes before expulsive efforts arise; while in the other the fœtus is not affected until the contractions have continued long enough to partially separate the placental attachments. Whatever doubt, however, may exist in all this, there can be no question as to the injurious effect produced upon gestation by ulceration or inflammation of the cervix uteri. Mr. Whitehead, of Manchester, England, has written a book, full of information, almost solely to illustrate this consequence of uterine inflammation.

Effect upon Labor.—The effect which inflammation of the os and cervix uteri exerts upon labor is not so apparent as upon the progress of gestation. Although I have watched patients. whom I knew to be laboring under inflammation of the neck of the uterus in parturition, I have not been able to perceive any increase in suffering or tediousness.

Even when induration and hypertrophy were both of several years' standing, no ill effects from them, so far as I could see, attended labor either at full term or prematurely. I have observed cases of abortion occurring in such patients quite as readily and without more troublesome symptoms than in one whose uterus was healthy. The general tissual changes going on in the uterus would lead us to expect this in advanced pregnancy, but I confess to some astonishment at having seen kindly, rapid, and complete dilatation in abortion at the early periods. It is equally singular to see the return of the induration after the involution of the uterus is fairly completed. One would suppose that the softening accompanying pregnancy would be permanent, and this is the case with indurations of recent date. I have not observed in such cases that the abortions were attended with more hemorrhage, or were more tedious or painful than when they occur as the result of some transient cause.

Effects upon the Post-partum Condition.—Of its effects upon the childbed, or post-partum condition, a favorable opinion cannot be given from my observation. A good getting-up is not to be expected with much confidence in patients affected with uterine disease. The most common effects in childbed is retardation of the processes of involution. The congestion consequent upon labor is protracted, the uterus remains larger and more sensitive than is usual, so that instead of the organ recurring to its primitive dimensions and susceptibility in one month, two or more may be required. The lochia, instead of subsiding in fourteen or twenty-one days, continues for weeks, or even months after it should have subsided; and when it goes off, it is apt to merge imperceptibly into leucorrhœa, which becomes persistent. Inability to walk or stand without great distress is the effect of the size and sensitiveness of the organ. A sense of bearing down, or of weight in the pelvis, pain in the sacrum, down the sciatic nerve or in the hip, harass the patient greatly, and these symptoms pass off so slowly that she is kept in bed an unusual length of time. Acute metritis not unfrequently supervenes, or acute inflammation of the cellular tissue at the side of the uterus. Phlebitis, pyæmia, and phlegmasia dolens are more likely to arise in patients who have chronic inflammation of the cervix.

On the other hand, it is a fact that these subsequent acute inflammations sometimes operate very favorably upon the cervical inflammations. Instances are not uncommon of patients being entirely cured of ulceration by the effects of gestation and labor upon the tissue of the neck and its mucous membrane. We are to hope for this favorable result only as a remote probability, because, as already stated, the condition of the parts is generally left in *statu quo*, or, if any difference is perceptible, it consists in an aggravation of the disease, and the patients get up from childbed rather worse than better.

CHAPTER IV.

ETIOLOGY.

Sexual Indulgence.—The unnatural social habits of civilized woman, and the circumstances which surround her, render her extremely susceptible to uterine disease. Coition, indulged in by the lower animals periodically, and only for the purposes of generation, at long intervals, is resorted to by man as the most common indulgence of his lower nature. The continued and extreme excitement in the sexual system ruins many of both sexes, but it produces the most disastrous effects upon woman, for obvious physiological reasons.

Improper Reading.—Less powerful but still efficient sexual excitement is found in the influence of lascivious books, so generally read by the young, as well as in the nature of the associations connected with most of the amusements of society.

All this is aided by heated rooms, stimulating diet, improper clothing, &c. At or near the periods of menstrual congestion these excitements operate with much more efficiency than at any other time.

Cold.—During the menstrual congestion, the application of cold to a large portion of the surface is also a fruitful source of uterine inflammation in very young girls.

Constipation.—Chronic and obstinate constipation keeps up a predisposing uterine congestion, and I have long since been led to regard continued constipation as a condition the most deleterious to female health.

Standing.—Constant standing also produces much evil; it is much worse than walking, or even than going up and down stairs.

Abdominal Supporters, Pessaries, &c.—Pressure upon the abdomen by miscalled uterine supporters, the improper use of pessaries, sponges, &c., may be enumerated as causes in certain cases of uterine inflammation. There can be no doubt, also, that prolapse of the uterus, when considerable, and other displacements, are sometimes the cause of inflammation of that organ. This, however, is a rare occurrence, as I think displacements are much more frequently the effect than the cause. Circumstances occur which may mislead us, if we are not careful, as to the proper relation between displacement and inflammation.

Severe Exertion, Jolts, &c.—We not unfrequently meet with patients who tell us that they were "perfectly well" up to the time of some severe exertion, jolt, or lift, when suddenly they felt something give way in the lower part of the abdomen, succeeded by pain in the back, hips, loins, groin, accompanied by a sense of prolapse and weight upon the perineum. Soreness and great permanent inconvenience persist thereafter, until the case becomes chronic. In such cases, the patient dates the beginning, not only of her trouble, but her disease, from the strain or jolt; and believes it to be the whole cause of her disease. A critical inquiry into the history of the case will generally convince us, however, that inflammation had preceded the accident, and that the uterus was probably rendered susceptible of the sudden depression by its increased size and weight. However this may be, the inflammation is greatly aggravated, if not originated by the circumstance.

Hemorrhoids.—The turgidity of the pelvic vessels kept up by hemorrhoids, prolapse of the rectum, vagina or bladder, or inflammation of any of these organs, must contribute largely to swell the number of uterine cases.

Pregnancy.—Although pregnancy is a physiological condition, and, in the nature of things, ought not to even predispose to disease of the uterus, recent investigation seems to indicate it as a prolific cause of ulceration. Dr. Cazeaux and other French obstetricians have examined a large number of cases of pregnancy, with a view to determine the frequency of ulceration in this condition; and having found ulceration almost always present, they have determined that the leucorrhœa of pregnancy caused ulceration of the uterine mucous membrane! As well might we expect to see ulceration of the bladder in consequence of diabetes mellitus, or ulceration of the skin in diaphoresis. Inflammation undoubtedly has the effect in this case, as in all others, of giving rise to the profuse and perverse secretion of mucus as well as the ulceration. There is no doubt but that in consequence of the dependent position of the uterus in its relations to the vascular system, it is more liable to congestions of both a transient and persistent character than any other viscus, not even excepting the rectum. These congestions are the predisposing conditions of inflammations generally, and if they are persistent and long continued, they excite, as well as predispose to inflammations. Constipation, standing on the feet for a long time, tight dressing, &c., act by impeding the upward tendency of the blood, causing it to leave the pelvis tardily, and thus keep hyperæmia in the uterine vessels until organic disease occurs. I cannot but believe that anything which will keep up these congestions for a sufficient time will bring about inflammation of the uterus in some part.

Abortions.—Abortions are both the cause and effect of inflammations of the uterus. It is hardly necessary to point out the deleterious effects of abortions produced by violence.

Abortion from Violence.—All the circumstances exist that are required. The violence is nearly always sufficient of itself at once to give rise to more or less acute disease. In cases occurring from other causes than intentional, or accidental violence, there are many efficient causes of congestion and inflammation. Probably one cause not usually thought of, is the too early assumption of the erect posture.

Bad Management after.—Being nothing but an abortion at an early period, it is not considered important by the physician that the patient keep the horizontal position; the patient sits up, walks about, &c., and the congestion existing continues sufficiently long to produce inflammation. Now, I think it is quite as necessary for the patient to remain quiet, in bed, until involution is well advanced, in cases of abortion, as in labor at full term. Many of the conditions inducing inflammation in cases of abortion are the same as arise in parturition. I shall, therefore, speak of them under that head, and the intelligent reader will at once perceive them as they are brought forward.

Labor.—The uterus, at the time of labor, is predisposed to vascular disease, on account of its extremely vascular condition; when labor comes on, the excitement is tumultuously increased, the nervous susceptibilities are enhanced, while the forcible contraction of the muscular part of its composition greatly adds to its excitation. When we remember the powerful compression of the os and neck by the child's head in passing through them, and even the frequent lacerations to which the mouth is subjected, it is astonishing that nature is competent, under the circumstances, to so completely restore so many parturient women to their former condition of perfect uterine health.

Decomposition of Productions of Labor.—In addition to all these, however, there is generally more or less decomposition of organic matter in the vagina, near the os and neck, giving rise to irritant products which, without proper cleanliness, might remain long enough in contact with the highly sensitive parts to cause inflammation. I know that nature should, and may, in most instances, safely be trusted to repair all the damage done in these ways when other circumstances are favorable; but these favorable circumstances are often wanting. The erect posture is too early assumed in many women, on account of their necessitous condition, or thoughtlessness and ignorance. This prolongs congestion of the dependent uterus, arrests or retards involution, and excites the uterus to inflammation; this inflammation is often prolonged by the continuance of the same cause until it becomes a fixed condition. The number of circumstances which cause and increase inflammation, in cases of parturition particularly, will be seen and understood without dwelling further upon them. We should remember them, and give our best care to patients passing through the conditions of the lying-in month, and thus avoid much suffering.

Vaginitis.—Inflammation originating in the vagina often spreads to the neck of the uterus, and occupies its mucous membrane externally, passes into the cavity of the cervix, and often, I think, to the cavity of the body of the uterus.

Gonorrhœa.—Gonorrhœal vaginitis is very prone to do so, and if not arrested, while yet in the vagina, and that soon after its commencement, the neck of the uterus is seldom left without permanent damage; and after gonorrhœal vaginitis is cured, it is frequently the case that the cavity of the neck is left inflamed. This may, and I think generally does, become chronic, unless removed by appropriate applications made directly to the membrane.

There is reason, too, for believing that the vaginal inflammation, in which profuse leucorrhœal discharges originate, arising from other than contagious causes, may pursue the same upward course, and leave behind the same grave chronic difficulties. It is well known that vaginal discharges are sometimes the result of general conditions, such as the scrofulous, for instance, so that we may have scrofulous vaginitis, and this may spread to the mucous membrane of the genital canal. Vaginitis may also arise from immoderate coition, masturbation, or the introduction of foreign bodies from perverse habits. What I have said above of the effect of vaginitis in causing cervical inflammation of the uterus, was intended to apply more particularly to this disease occurring in adults; but there is another condition under which it occurs, that I think has escaped the attention of medical men, or at least has not attracted sufficient notice, viz., the vaginitis of children.

Vaginitis of Children.—I think I have observed several instances in which, before the appearance of the menses, the cavity of the cervix must have been affected with inflammation extending from the vagina. Indeed, if the history of patients who very early commence to complain of signs of inflammation of the cervix be properly traced, it will be nearly, if not always, found that they were to some extent the subjects of leucorrhœal discharge during their childhood. The kind and locality of the disease arising from infantile vaginitis is almost peculiar. It is situated inside the cavity of the neck, and if the os uteri is examined with the speculum, when the disease is not great there will be found but little, if any, unnatural appearance, save the issuing of muco-pus from it. The os is often contracted in size; it is very seldom enlarged.

These young patients do not generally complain much of suffering from their vaginal inflammation until the commencement of their menstrual visitation, when they have severe pain at each time. The suffering ordinarily increases as the functional activity of the uterus increases, until the patient is a confirmed sufferer with dysmenorrhœa or menorrhagia. At other times, instead of having much direct uterine suffering, the general nervous system is most affected, or the vascular or nutritive systems become seriously deranged at the period when the menses should appear. It is not the usual opinion, but I am, nevertheless, inclined to the belief that chlorosis and chorea are sometimes the effect of derangements thus produced.

The above short and imperfect sketch of the causes of inflammation of the mucous membrane of the uterus will give but an inadequate idea of the vast number of causes which produce the inflammation in question. There is no mucous cavity in the body that is subject to so many causes of intense excitement, arising from the nature of its functions, from its accidents and abuses, as is that of the cavity of the female genital canal. Hence it is not wonderful that this cavity is very much more frequently the seat of disease than any other mucous cavity in the human body.

CHAPTER V.

PROGNOSIS.

A just estimate of the chances of making a cure, or of spontaneous recovery, or, in other words, correct notions of the prognosis of a disease in any given case, has necessarily great influence upon our treatment; and a correct prediction of the progress of a case, or of its ultimate result, has an important relation to our reputation and to the confidence of our patient. It is especially important to be able to give a reliable prognosis in cases in which the profession as well as the patients are not perfectly satisfied about the pathology and therapeutics in reference to them. Too unfavorable an opinion discourages our patient, and precludes us from having a fair opportunity of exercising our efforts; too favorable an opinion, one not justifiable by the result, brings disappointment to the patient, injures the reputation of the practitioner and the profession, and is also apt to influence improperly the inexperienced medical man against the treatment adopted. The general principle that should govern our prognosis is temperance. We should temperately encourage our patient, if we can conscientiously do so, and if our judgment will not allow us to do this, we should express, temperately and cautiously, an unfavorable prognosis; and hope should never be extinguished until the patient is moribund. Too many good reasons will suggest themselves for this last course to require any argument in support of it. What I have said of a guarded prognosis, and the necessity of not giving a sweeping and absolute opinion, seems to me peculiarly applicable to the diseases of which I am now treating. Physicians have not all been convinced of the propriety of treating uterine diseases with the speculum; a large number are entirely, and conscientiously, opposed to it. They are made so, undoubtedly, by the failure of local treatment to fulfil the hopes originated by its most ardent advocates. It does not do what they are told it will do; it certainly does not in all cases. The only grave error I think committed by that benefactor of womankind, Dr. Bennett, in his work on the Unimpregnated Uterus, is that his book leads his readers to believe that he scarcely, if ever, fails to cure his cases. This is the impression made upon most physicians who read his book. However true it may be, with reference to the practice of so able a master, I think it would be an unjustifiable expectation on the part of the profession at large. From what I have heard and read of the opposition of medical men to local treatment in uterine disease, I think this unrealized expectation of success from local treatment, is one of the main causes of it. Upon trial, medical practitioners become disappointed with the results as they were led to expect them, and abandon the plan as a failure. While I cannot coincide with Dr. Bennett as to the almost universal success of local treatment for uterine inflammation, I am of the opinion that it is greatly superior to any other with which I am acquainted. Prognosis must depend for its reliability, to some extent at least, upon a correct and complete diagnosis of the whole condition of the patient.

Uncomplicated Case Favorable.—The probability of recovery of health will depend upon the absence of any important general diseases in conjunction with the local. We should remember that the patient aims at recovery of health, instead of merely the cure of any one part of the ailments. An important matter is to determine the pelvic complications, if any exist, and how far they are curable, before we pronounce a prognosis.

Prognosis without Treatment.—What is likely to be the progress and result of the disease when allowed to go on without interference? Generally, it will go on from bad to worse. This is particularly the case with the childbearing woman; it is almost equally true of the menstruating unmarried woman. In the latter, however, if she avoids the causes which aggravate it, she may not get worse; but if her situation, or her inclination, subjects her to the aggravating causes, she will also become worse. Not unfrequently the patient recovers after the "change of life" takes place. The cessation of the menstrual congestions, if other things are favorable, seems to determine a gradual recovery.

This I fear, however, is far from being as frequently the case as we might suppose from reading, and judging physiologically of the matter. Indeed, some of the most obstinate cases I have met with were patients in whom the disease had outlasted the change of life.

Not often directly Fatal.—Notwithstanding the tendency of the disease to get worse during the whole menstrual life of the patient, and to subside only with the subsidence of uterine activity, it seldom proves fatal directly. Nor do the most common and immediate effects of it proceed to a fatal extent. The debility, the imperfect or perverted hæmatosis, or the nervous energy, seldom becomes so great as to be the immediate cause of death. This, however, sometimes does occur, and we should indulge a false security to suppose that our patient could not thus die. I think I have seen more than one instance of death thus resulting. The nervous and muscular systems very rarely become so influenced by perverted innervation and hæmatosis as to assume dangerous or even fatal complicating conditions.

Indirectly Fatal.—As very correctly stated by Dr. Bennett, such an unnatural condition of the nervous system and blood is engendered by the disease as to destroy the capacity of the patient to resist or ward off the attacks of the acute diseases to which she may be exposed, or the chronic ones for which she may inherit a strong predisposition. It is difficult also to resist the belief, although I have not verified it by observation, that puerperal fevers and post-partum affections are more likely to occur and assume a dangerous or fatal state in patients affected with chronic uterine disease. I need hardly mention the increased hazard to married women from abortions, and the diseases intercurrent with them.

Prognosis in different Varieties.—There is some difference, other things being equal, in the gravity or seriousness of different varieties of inflammation. Some produce much worse effects upon the constitution, are more obstinate and protracted in their duration, and even resist treatment with greater persistence than others. When the inflammation is confined to the mucous membrane outside the os uteri, the prognosis is most favorable; if the inflammation exist in the mucous membrane of the cavity of the cervical canal, it will be more obstinate and difficult to eradicate; and some of these cases are exceedingly so. In the class of cases in which the inflammation has extended to the submucous tissue, the prognosis, so far as it respects a perfect cure is concerned, is unfavorable; it becomes especially unfavorable when the inflammation has lasted so long as to materially alter the shape, size, and consistency, by deposition of fibrin, of the neck of the uterus. In these cases the inflammation is not all that has to be encountered; but the organic alteration must be corrected. This cannot always be perfectly done. If the neck of the uterus is indurated, enlarged, and nodulated, we can only partially restore the organ to its original softness, evenness, and size. And to do this, requires a long time and patient and judicious management.

Prognosis under Treatment.—In cases of uterine inflammation and ulceration in general, what is the prospect under properly conducted treatment? The prospect of cure is comparatively favorable. I mean by this statement that, compared with other diseases which produce as much suffering, the prognosis, under proper treatment, is quite favorable. What the percentage of cures would be if summed up, I could not say; but it is large. A more circumstantial consideration of the prognosis I think would be profitable. With reference simply to recovery or death, the prognosis is favorable, because, even when a cure is not effected, as we have seen, it is not usually fatal.

Can the Inflammation be always removed?—And if removed, will the local symptoms always subside? The local inflammation can nearly always be removed; but with its removal, the local symptoms do not always leave the patient.

Will the several Symptoms always subside?—The inflammation, so far as we can see, may generally be removed; but many of the symptoms, as the pain in the back, groin, or elsewhere, may persist, to the great discomfort of the patient. I have endeavored to show that many of the symptoms depend upon the congestion kept up in the whole of the pelvic organs; and that these congestions are not unlike those produced by the menstrual molimen, and that this persistent congestion depends upon the presence of the inflammation in the cervix and os uteri. This congestion sometimes outlasts the inflammation, and thus keeps up some of the local symptoms. But by far the most frequent reason why the local symptoms do not subside, is the persistence of inflammation to some extent. This may be out of sight, and consequently undiscovered; but if it is mucous inflammation, we may know that it is not cured while there is a superabundance of mucous secretions or vitiated mucus or pus in view. If it is submucous inflammation which still exists there will be unnatural tenderness when touched by the finger or instruments. This tenderness being unnatural, would indicate still some inflammation.

How long will it take to cure the Inflammation?—In what length of time can we reasonably expect a cure of the local inflammation? No certain answer as to this can be given from any mere observation of the case in the beginning. From three to twelve months or several years should be the latitude given in most instances. A shorter time than three months is uncommon; and we might in many instances not reasonably expect a cure in twelve. In order to fulfil the expectations of the patient and of ourselves, we should take plenty of time, and we should not lead her positively to expect a removal of all the symptoms when the treatment has terminated; for they sometimes subside so slowly that they continue many months after the treatment has ceased. The general or sympathetic symptoms sometimes become a sort of habit, and continue after the disease which called them into existence has been cured. This is particularly the case with the great degree of general nervousness which renders some patients miserable; but in the great majority of cases they do subside very readily as the ulceration is cured. When they do not, judicious treatment directed to them will do more for them, after the ulceration is removed, and will almost invariably relieve the system of them. For the removal of these general symptoms, time is an item of the utmost importance, and we do not do justice to our own reputation or to the patient, by fixing the time too positively in which relief may be expected.

Prognosis influenced by Age of Patient.—The age of the patient, I have thought, had a good deal to do with the readiness and completeness of recovery from all the troubles of uterine disease. Young women will recover quicker than the old; the naturally robust and active woman sooner than the delicate and inactive. If there is an hereditary predisposition to insanity, or any general nervous disease, there is great likelihood of its being excited into activity by uterine irritation; and when once started, it is likely to assume a permanent and durable form. We should not, therefore, promise too much, in patients whose general health has been long seriously affected, as it is impossible to predict the measure of benefit to be derived from a removal of the cause of the general affection in the cure of the local.

How and when does Relief come in Favorable Cases?—But, in cases in which relief from the general and local symptoms readily succeeds the treatment, there is considerable difference as to the mode in which the relief comes. In very many cases the patient experiences benefit from the beginning. In the first month she feels the cessation, or a great amelioration of some of her symptoms, generally of the local pains, and she continues to improve until entirely cured. In other instances, the symptoms are aggravated for several weeks, and there is no improvement until after the local treatment is discontinued; again, relief does not follow for some months, and yet by judicious general management it is secured. In a great majority of cases, we may very plainly see the beneficial effect of our treatment; if not before, certainly by the time we have procured the complete resolution of the local disease. As I have before intimated, the general sympathetic effects are sometimes kept up by local complications, and will subside only when they are removed.

Will the Functions be restored?—An important part of prognosis, one in which our patient often feels a deep interest, is the determination of the prospect of restoring the functional derangements of the uterus. As it has been before stated, inflammation and ulceration of the cervix uteri often cause sterility. This condition occurs under two different sets of circumstances: in one, the patient never conceives after marriage, and may remain sterile during her whole lifetime; in the other, she conceives and miscarries, or even goes to full term of pregnancy for one or more times; and then as the inflammation and ulceration become established, she ceases to become pregnant. Where the patient has been married for several years, and does not become pregnant, the cure of the disease is not generally followed by productiveness; and when it is, it is usually after the lapse of a long time, sometimes amounting to several years. Although this is the most common condition of the functions, sometimes, after treatment fertility is the immediate effect of a cure. I have noticed that patients who remain sterile in this way, usually have very scanty menstruation. And we cannot reasonably hope that our patient will cease to be sterile when she is cured, if this very scanty menstruation is not, or cannot be corrected. I am inclined to think that in cases of this kind the ovaries are in some way, probably by chronic inflammation, also rendered unfit for their duties. Those who are restored to fertility by the cure of the inflammation, always, or nearly always, have a normal condition of the menstrual flow.

Patients who have conceived and miscarried, or borne children, but become sterile, are usually cured of their sterility with the cure of the disease of the uterus. Yet repeated instances have come under my observation, where a miscarriage soon after marriage has resulted in permanent and incurable sterility. Most of all these cases were abortions induced by forcible means. The inflammation seemed sufficiently intense to destroy the capacity of the uterus for lodgement of the fœtus; or, at any rate, to render that organ in some manner unfit for the discharge of its part of the function of generation. If a woman has had several miscarriages, or borne a number of children, and then becomes sterile, there is great reason to hope that she will at once become fruitful as soon as the inflammation is cured. This result will be the more likely if menstruation retains its natural characteristics. The habit of miscarrying is generally quite effectually broken up by the cure of the disease on which it depended, so that we may pretty confidently assure our patient that as soon as the inflammation is cured, pregnancy will go on uninterruptedly to the full term. We should, however, promise this only of future pregnancies; as, according to my observation, a cure undertaken during the existence of this condition is not very promising, although we have good authority for making the attempt. I am not satisfied that the attempt is always best to be made, and generally wait until pregnancy is over, and even stop the treatment if I have begun. I could cite many cases corroborative of the statement that habitual abortion is cured by the relief of the inflammation. It will not be amiss to state the result of my observation as to restoration from the menstrual deviations which attend, and for the most part depend upon the diseased uterus. Generally, this restoration takes place, but it certainly does not always. I think different sorts of menstrual trouble are differently influenced by the cure of the inflammation. Scanty menstruation often remains permanent after the cure of the diseased cervix, and much more frequently resists treatment than any other derangement. Where the menses have been wholly suppressed, we may hope for better results from judicious management. In fact, the stimulation of the uterus generally restores this function when absent on account of chronic inflammation, unless, as is sometimes the case, so much organic alteration has been brought about as to destroy to some extent the texture of the organ. Menorrhagia often continues with considerable obstinacy after all the disease of the cervix is removed; but it is nearly always much moderated, and quite frequently entirely cured, and ceases to trouble the patient before the inflammation has wholly disappeared. Where it is obstinate, it will nearly always be found to be the case that after the lapse of a few months it begins to improve, and after a while the menstrual discharge will not exceed the natural quantity. I think we may pretty confidently hope that by the exercise of a little patience we will cure this functional disorder of the uterus, where it has depended upon inflammation of the cervix.

Dysmenorrhœa, when dependent upon this cause, disappears often very readily under the influence of treatment directed to the cervix, but we should be careful to distinguish between it and that which depends upon other causes. Very commonly one of the first good effects of local treatment is to ameliorate the suffering during the menstrual discharge. This is often remarkably the case, the first menstrual effort being so much better as to astonish the patient and her friends. It would hardly be justifiable, however, to promise, generally, such ready relief; for sometimes this feature of the case remains quite obstinate, and causes the patient a great deal of suffering after the inflammation is entirely cured.

Complicated with Phthisis.—In the course of my practice, it has occurred to me to have cases complicated with tuberculous disease of the lungs, and some of these patients have seemed to run down more rapidly after their recovery from the uterine disease than before, on account of their softening and discharge. I have not had an opportunity to observe a sufficient number of such cases with that scrutiny so necessary to arrive at a correct conclusion. It might be supposed that, on account of the derivative influence of the uterine disease, the consumption was kept in abeyance by its continued existence. On the other hand, the debilitating effects upon the system at large, which it undoubtedly exerts, might with equal propriety be expected to co-operate in the general prostration.

Throat Disease.—The frequent complication and the effects exerted by the one upon the other of throat affections—pharyngitis and laryngitis—and uterine affections, makes it a matter of interest to determine what, if any, is the effect upon the diseases of the throat, of curing ulceration of the cervix uteri. This may be regarded by most of my readers, and probably is an irrelevant question in this connection, but I think careful attention to it will lead to a different way of thinking about it. I am persuaded that some, at least, of the chronic sore-throats of this climate can be much more easily cured after the uterine complication is removed. Women often believe that there is an intimate connection between them, and hope that the cure of inflammation of the uterus will relieve the throat; and I have seen cases in which I was almost ready to believe there was some encouragement for the opinion.

Skin Disease.—Psoriasis, lepra, and some other of the chronic forms of scaly eruptions, coexisting with inflammation of the cervix uteri, have been aggravated or ameliorated as the uterus grew better or worse; when the uterus is better the eruption is worse, and the converse. I have noticed several cases in which this seemed to be unequivocally true; and it is remarkably the case in two patients now under my observation. Without my speaking of it, they have both remarked it. If this observation should prove true on a large scale, it would indicate the extension, in a modified form, of this chronic inflammation to the mucous membranes, and afford us a valuable hint for the appropriate mode of managing a class of very obstinate cases. The above facts have an important and direct bearing upon our prognosis; for, according to my experience, the cases attended with these chronic skin diseases are very obstinate and protracted.

Cure remains Permanent.—When cervical uterine inflammations are once cured, are they likely to return? It is a popular belief that these uterine diseases cannot be permanently cured; that they will keep returning. This belief is, no doubt, supported by the fact that many of our patients are constantly laboring under the causes that originally produced the affection; and, therefore, are likely to have them reproduced. Of course, patients thus situated will have a return of the diseases, but where the causes can be avoided and the cure completed, there is no reason why they will not remain cured with as much certainty as any other disease susceptible of perfect removal. I cannot refrain from here expressing the opinion, however, that a large majority of the cases that thus thwart our hopes never are entirely cured, and I believe great discouragement arises from want of the experience necessary to decide when the disease is entirely removed. I have met a number of instances in which the practitioner supposed he had removed the inflammation, but the symptoms remained; where an examination revealed a discharge of muco-pus from the mouth, showing inflammation still remaining inside the neck, and discoverable only by the discharge.

CHAPTER VI.

COMPLICATIONS OF INFLAMMATION OF CERVIX.

Complications.—Various and troublesome intra-pelvic complications are often observed in connection with uterine disease. These complications for the most part arise during the existence, and generally as the effect, of the disease of the uterus; they may, of course, also arise as independent affections. Notwithstanding the frequent secondary origin of these complications, after they have continued for a considerable length of time some of them become permanent, and after the originating disease has subsided they go on indefinitely if not cured.

Vaginitis.—Probably the most common of them is vaginitis, in some form; ordinarily in that of erythematous inflammation of the mucous membrane, which is indicated by an increased mucous discharge, some tenderness and heat. Instead of the inflammation being thus moderate, there may be copious muco-purulent discharge, great irritation, and so much tenderness as to render an instrumental examination very painful, and often unsatisfactory. Such severity of inflammation is apt to be of short duration, and dependent upon some superadded cause of the inflammation. The inflammation is usually more moderate and persistent, continuing more or less for weeks or months together. Another form of complicating vaginitis is eruptive, and, although not usual, it yet sometimes accompanies the simple variety. The eruption in the milder form is vesicular. Small vesicles appear somewhat thickly studding the inner surface of the labia, on the nympha, the membrane of the vestibule, and sometimes the cutaneous surface on the edges of the labia majora and the anterior edge of the perineum. This eruption is attended with great heat, or a burning sensation, and not unfrequently with intolerable itching. The vesicles are not very thickly set upon the surface, but the latter is of a fiery red color. A greater or less amount of serous discharge keeps the parts wet and sticky. Almost always this mild eruptive variety is paroxysmal, and generally appears simultaneously with the commencement of the menstrual discharge, and has seemed to me to be dependent upon the acrid discharge accompanying it, and the congestion present at such times. The eruptive variety of vaginitis is sometimes much more severe in grade, and the vesicles are changed to pustules, and the accompanying inflammation much greater. Fortunately this is not nearly so common as the first two forms. Vaginitis sometimes has its origin, I have no doubt, in an extension of the mucous inflammation from the neck; but frequently, I think, the inflammation is caused by the acrid, irritating nature of the perverted secretions from the mucous membrane of the cervix and by want of proper cleanliness.

Urethral and Cystic Inflammation.—Urethral and cystic inflammation also not unfrequently result from or accompany cervical inflammation. It is not necessary to give their symptoms in detail. The main fact to which I desire to give expression is, that when there are symptoms of cystitis or urethritis we should be watchful for the probable occurrence of inflammation of the bladder and urethra, and be aware of the importance of giving attention to them as complicating diseases. For I think I have seen indubitable instances of cystitis and urethritis which could be traced to this cause, continuing after the uterine disease was cured. When not properly attended to, they may induce nephritis. The inflammation of the neck no doubt directly induces inflammation of the bladder, by reason of its immediate apposition to its walls; and while this inflammation ordinarily is of short duration, yet it sometimes becomes very persistent, and even permanent. The attacks, when acute in grade, as they sometimes are, become extremely distressing, and absorb the whole attention of the patient, and demand the prompt interference of the medical attendant. More commonly the grade of inflammation is mild, and confined to the mucous membrane of the organ. The scalding micturition indicative of urethritis is often distressing to a great degree, and is not unfrequently very persistent. This urethritis and cystitis, I think, are caused by migrating inflammation from the vagina in some cases, and the inflammation probably goes on through the ureters to the pelvis of the kidneys. When cystic inflammation is persistent and somewhat severe, it often passes for the main disease. The symptoms of cervical inflammation of the uterus being overwhelmed and obscured by the more urgent and distressing vesical affections, is not thought to be the origin of the trouble. Although the vesical symptoms, as before stated, may become urgent, and the inflammation assume an important prominence in the case, usually this complicating affection is slight, and manifested by very mild and transient symptoms. In this form, cystitis and urethritis are very common indeed.

Cellulitis.—A more formidable, troublesome, and perplexing complication, however, is a chronic or subacute form of cellulitis, as it has been named by Prof. Simpson. It consists of inflammation and suppuration of the cellular tissue contained in the duplication of the peritoneum, at the side of the uterus. I think this is a frequent complication, and more frequent, according to my observation, than we are led to believe by any description I have ever met with. When it is present, it embarrasses our diagnosis and should very materially modify our prognosis. I have met with instances in which it remained unnoticed, and exercised a very embarrassing effect upon the treatment and the progress of the case for a long time. This complication is important for two main reasons at least, viz.: 1st, the great obstinacy with which in the chronic form it resists treatment; and, 2dly, from the fact that the pelvic or uterine symptoms do not subside while it lasts, even when the uterine disease is removed. It is likely to occur in two forms, differing considerably in intensity and duration; the acute and the chronic.

Acute Cellulitis.—In the acute form the symptoms are violent, and run their course somewhat rapidly. The patient, after some exposure, or more than ordinary exertion, experiences a great increase of pain in the pelvis; it usually occurs on one side, and rigors supervene, which are succeeded by febrile reaction of high grade. The pain is constant and often excruciatingly severe, of a tense and aching character. It is sometimes attended by paroxysmal exacerbation, but it is generally free from it. The fever, pain, and great soreness, continue from six to twenty days, or even longer. The fever gradually becomes more remittent, and finally intermittent, being terminated, or nearly so, every night by copious perspirations. The pain continues, however, until it is relieved by a discharge of pus per vaginam, rectum, or urethra.

Suppuration in Cellular Tissue.—If the discharge is free and copious, immediate and almost complete relief follows; if, as is much more frequently the case, the discharge is small, the relief is only partial, and the patient lingers in a state of great suffering for weeks, and even months, before the discharge is completely effected and the cavity of the abscess filled up. During the existence of these acute symptoms, if we examine per vaginam with the finger, we will find the mucous membrane hot and exceedingly tender to the touch.

Diagnosis of Cellulitis.—In seeking to ascertain the relation of the organs, the uterus generally will be discovered situated near one side of the pelvis, and fixed in its position, so that it cannot be easily moved in any direction; or it may be in the middle of the pelvis, and a little lower down upon the perineum. When it is to one side, we may feel on each side of it solid tumefaction, filling up to a considerable extent, if not completely, the lateral and anterior portions of the pelvis; and if we press upon this hard and tumefied part, we shall cause great complaint of tenderness. The patient will cry out with the pain produced by it. If the uterus is central in its position, the hardness, pain, and swelling, will occupy one side of the pelvis, and while it will give the patient great pain to carry the finger up the side of the uterus, where this tumefaction is situated, on the other side there will be no tenderness.

Extent of Cellulitis.—These inflammations invade the cellular tissue in the pelvis to a greater or less extent in different cases, and sometimes the infiltration is so great as almost wholly to fill up the cavity of the pelvis. In other cases there is only a very small amount of induration, not larger than the thumb. Now, attacks of the kind above described cannot deceive a careful practitioner; but the milder and less pronounced variety may go unnoticed unless we are watching for it.

Chronic Cellulitis.—The patient in the milder form is seized with some increase of pain in the back or groin, or elsewhere about the pelvis, which lasts for three, four, or five days; and after a discharge of very little pus, it subsides, and leaves the patient in her usual condition. This mild form may, and indeed often does, occur as an original condition, but much more frequently it follows at some distance of time an acute attack, such as I have above described. However this may be, it nearly always represents a small nidus of chronic inflammation by the side of the neck of the uterus. The chronically inflamed cellular tissue in this region is not so great in amount as to cause any febrile excitement, and in fact attracts but little attention, except when it is aggravated into the suppurative process from time to time. I have met with instances in which suppuration and discharge of pus from a small chronically inflamed point of cellular tissue had recurred every few weeks for twenty or more years. And it is often the case that patients having this inflammation will experience exacerbations every month, before or after menstruation, and thus these attacks may pass for cases of dysmenorrhœa. The frequent discharges of pus with these slightly painful exacerbations should cause us to make an examination, when we may generally find a point of induration and tenderness. The results of the examination will be most satisfactory at the time of the exacerbation, as the parts will be more tender and the swelling greater.

It is not necessary for me here to go any further in the description of this intra-pelvic abscess, as I only wish to call attention to the fact that it is not an unfrequent complication of uterine disease; that the symptoms attending it very much resemble inflammation of the neck of the uterus; and that when it continues after the inflammation and ulceration of the neck are cured, the uterine symptoms do not subside as readily as when these last are not thus complicated. To the inexperienced practitioner it is a troublesome and perplexing complication, and if not particularly cautious he is betrayed into an unjustifiable prognosis, if nothing worse. Intra-pelvic inflammations of this kind, although occasionally independent and uncomplicated, I think are much more frequently associated with chronic inflammation of the uterus. And I cannot but determine, as the result of my own observation, that they are secondary to the uterine inflammation in a large majority of cases, and caused by an extension of it.

Cause of Cellulitis.—Dr. Bennett thinks increase of inflammation of the uterine tissue, produced by strong cauterization, occasionally the immediate cause of cellulitis. Although this is doubtless true, yet a great many cases occur in which no local treatment has ever been resorted to. I do not remember to have met with but one case in which this could have been the cause, and in that case it did not manifest itself until four weeks after the caustic potassa had been applied, for cervical induration and tumefaction. It is reasonable to suppose, however, that any circumstance which would excite the vessels as this does, might, and most likely would, enhance the probabilities of cellulitis.

Rectitis as a Complication.—The rectum is very often diseased in uterine cases; in fact, it is not often that inflammation of the uterus lasts for many months without affecting the rectum to a greater or less degree. Chronic inflammation of the rectum is quite a common complication with certain kinds of uterine diseases. The inflammation is evinced by the tenesmus, frequent discharges, and the increased secretion from the mucous surface. The symptoms are those usually present when the rectum is inflamed from any other cause. The degree of inflammation will cause quite a difference in the intensity of the symptoms. In very many instances there is moderate tenesmus, causing five or six stools in the twenty-four hours. These are partly fecal but thinner than natural, and loaded with mucus; or there may be more tenesmus, with more frequent efforts at stool, less discharge, which consists mostly of mucus, streaked with blood. The discharges from the rectum in bad cases may be more or less purulent in character, or may consist exclusively of blood.

Diagnosis of Rectitis.—Where there is rectitis, it is usually tolerably high in this organ, being two, three, or four inches above the anus; and in our examinations, if we press upon the rectum from the vagina, it is found to be quite tender to the touch, and always empty. It is too irritable to retain fæces for any length of time. So that when we find a mass of hardened fæces occupying the rectum, very perceptible through the posterior wall of the vagina (and we will often find such), we may be pretty sure that the rectum is not much affected.

Stricture of the Rectum.—Another condition of the rectum

which is apt to be associated with rectitis in uterine disease, is stricture of this organ; the stricture varying, of course, as to the time it has lasted and the severity of the cause.

Fistula in Ano.—They may both be succeeded and accompanied with fistula in ano. These complications have their own symptoms, and must be investigated and treated as though they were independent affections, while we attempt to remove the cause.

Causes of the Rectal Diseases.—How these three different forms of rectal disease are produced by the disease of the uterus, although not very plain, may, I think, be generally explained. The inflammation doubtless extends from one tissue to another in rare instances, but more frequently, I think, it is caused by the pressure of the uterus upon the rectum. The rectum lies on the sacrum; and the uterus often becomes so heavy that its supports are not sufficient to keep it in place; it settles upon the rectum and presses it against the hard surface of the sacrum, thus irritating it very much, bringing about congestion and inflammation first, spasmodic and then organic stricture, and subsequently ulceration and perforation of the mucous membrane of the rectum. The lumps of fæces or other substances burrow through the rectum in this ulceration, when suppuration and exulceration establish a fistulous opening.

Prolapse of the Rectum.—But without much inflammation the rectum is sometimes prolapsed so that it protrudes from the body, either through the anus or the ostium vaginæ. In long-standing cases of uterine disease, great relaxation of the mucous membrane of the rectum is a frequent occurrence; and then, in every effort at stool, it falls in large folds through the anus, often entangling the fæces in them, so that the patient is under the necessity of picking them out before the evacuation can be completed. Or, what is less frequently the case, as the tenesmus of defecation attempts the expulsion of the contents of the rectum, this organ is forced forward into the vaginal cavity, and externally between the labia, so as to form a tumor external to them with the fæces contained in it. The evacuation of fæces from the rectum is very difficult in this complication, and the patient will tell us that she is obliged to introduce her fingers into the vagina, pressing the whole mass backward and downward toward the opening of the intestine. It is astonishing to what extent such displacements of the rectum are carried. Its folds often protrude sufficiently to cause a tumor below the anus or external to the vagina large as a man's fist.

Hemorrhoids.—Hemorrhoids is another form of disease of the rectum and anus, which complicates diseases of the uterus. They of course will not require a distinct description; their frequent occurrence renders everybody familiar with them. The pain resulting from inflamed hemorrhoids often masks or simulates inflammation of the neck of the uterus; and when they are associated, the cure of either will not remove the symptoms, so that we need not be surprised at their greater obstinacy when they coexist. The prolapse of the rectum and hemorrhoids are the unquestionable results of uterine pressure. The continued congestion kept up in the rectal vessels by the constant pressure of the uterus upon the rectum, hypertrophies the mucous membrane, and causes varicosity of the extremities of the veins, and in this way induces both results. They are, therefore, the *indirect* results of inflammation of the uterus, this last bringing about a change in the position of the uterus, so that in some portions of it, it presses the rectum against the sacrum so firmly as to embarrass its circulation and cause the changes above described.

Hypertrophy of the Rectal Mucous Membrane.—The rectum is not only prolapsed, but the mucous membrane is hypertrophied quite largely, before it can appear externally; and in conjunction with this hypertrophy there is also great relaxation of the fibres of the rectum and sphincter ani, or of the fibres of the vaginal walls, to allow the escape of the parts to the enormous extent which sometimes takes place.

Displacements of the Uterus.—The most common displacement of the uterus, where these two last rectal complications are present, is the subsidence of it in the axis of the superior strait. This brings the neck of the uterus straight down upon the rectum, and the whole weight of the uterus rests upon it. This brings me to the consideration of the most frequent of all complicating circumstances connected with chronic inflammation of the uterus, viz., uterine displacements. So frequent are these displacements in this relation, that, as I have before stated, they are regarded as the causes of all the associated difficulties.

I cannot assent to this view of the subject, but believe them to be frequently, if not almost invariably, the effects of inflammation, and am confident they are most important and mischievous complications, and probably give rise to more suffering than any complicating condition whatever. As I have already stated, it is most frequently the displacements that cause stricture, hemorrhoids, and prolapse of the rectum. By the uterus being crowded down upon the rectum, these affections may be produced. It will not be expected that I shall dwell with any great degree of minuteness upon the different degrees or characters of displacements, or give a full description of them here, as I only wish now to speak of them as a complication of chronic inflammation of the uterus.

Subsidence of the Uterus.—The most common displacement I meet with is a subsidence, or lapse, of the organ, while its vertical axis remains what it was before the change of position. This does not bring the uterus, or any part of it, nearer the vaginal orifice; the lower end of it settles down upon the lower bone of the sacrum, while the fundus points upward toward the umbilicus. In examining per vaginam, instead of finding the os uteri upon, or nearly upon, a level with the inferior border of the symphysis pubis, and touched by introducing the finger almost directly backward, it is necessary to bend the finger over the upper edge of the perineum, and carry it back and downward to the lower end of the sacrum. This displacement is very frequent, according to my observation, and does more injury by pressing upon the rectum, and gives more distress, than almost any other displacement. It almost always obstructs the passage of the fœces through the rectum, and makes the patient feel as though the bowel was constricted at the point of pressure. After long continuance, it induces, in many instances, organic diseases of the rectum, inflammation attended by tenesmus, mucous and even bloody discharge, hemorrhoids, &c. All these rectal complications above-mentioned may arise in this way.

Anteversion.—The inflamed uterus is also anteverted, more or less, in many instances, so that the fundus presses heavily upon the bladder, while the os, higher up than in the first-named displacement, presses the rectum against the sacrum. But as most of the weight of the uterus is upon the bladder and anterior wall of the vagina, the rectum is not so distressed.

Pressure upon the Bladder.—The greatest inconvenience is felt on account of its pressure upon the bladder. Frequent micturition, sense of weight behind the pubis, &c., are its symptoms.

Retroversion.—Retroversion is also not unfrequent as a troublesome complication. As the fundus presses upon the lower part of the rectum and perineum, while the neck and os press upon the urethra and bladder, there is dysuria and rectal tenesmus, of greater or less intensity. The symptoms will be modified by the greater or less degree of malposition.

Prolapse.—Common prolapse, with the mouth following the axis of the vagina, is the least frequent of these displacements, as I have observed them. It sometimes occurs, however, to a very great extent, and produces a great deal of distress. Compared with the other forms of displacement, it produces less inconvenience when present in the same degree. It certainly does not interrupt the function of the other pelvic viscerae so much as subsidence, retroversion, or anteversion. Where excessive, it gives a sense of dragging and perineal tenesmus that are very disagreeable, but it does not cripple the patient, and render her unable to walk or stand, as is the case with the other displacements. While displacements aggravate the sufferings connected with diseases of the uterus, they render the treatment more difficult, and often imperfect, on account of the difficulty of exposing the os, and bringing the axis of the uterus to correspond with the direction of the speculum.

Theory of Displacements.—I cannot now enter into the theory of displacements as complications of inflammation. I believe they are one of the effects of the pre-existing inflammation; that they are brought about by the inflammation increasing the size and weight of the uterus, and thus causing it to settle down by virtue of its weight in spite of its supports; that the suffering caused by the displacement results from its pressure on morbidly susceptible organs, made so, perhaps, by a long continuance of the pressure, and by the sense of soreness in the inflamed uterus itself, and also in part by traction upon the lateral and round ligaments. Still, I have no question that in very rare instances the displacement results from other causes than inflammation, and then I can easily comprehend how it may produce inflammation in the uterus. The circulation must be embarrassed, congestions will readily occur on account of pressure and forcible flexion of the veins and arteries, and inflammation is very apt to follow long-continued congestion, &c.*

Flexions are very common complications of chronic inflam-

Fig. 1.

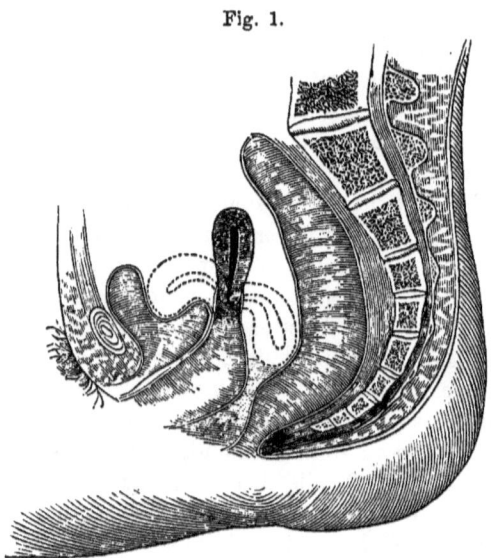

This figure shows the Uterus in proper position. The dotted lines anterior, its position in anteflexion; those posterior, in retroflexion.

mation of the unimpregnated uterus. Flexion and inflammation are so frequently associated, that I seldom see the flexion without the latter condition.

---

* See Displacements.

CHAPTER VII.

POSITION OF INFLAMMATION.

Submucous, or Fibro-Cellular Inflammation.—Chronic inflammation may originally attack any portion of the uterus, from the fundus to the lower extremity of the neck, and be seated in either the fibro-cellular or mucous tissues. The part of the organ most likely to be attacked, however, is the cervix, and of this the mucous tissue is nearly always the seat of disease. When the inflammation originates or invades the fibro-cellular tissue of the uterus, it is soon followed by enlargement of the portion inflamed. If inflammation affects the wall or substance of the cervix, the enlargement is the result of the infiltration of serum in the tissues. After the effusion has lasted for a time, the fluid portion of the effusion is absorbed and the fibrin is left behind, and the part is hard and irregular in shape as well as tumefied, and then we have a hard, tender tumefaction in that part of the uterus. When the substance of the cervix is chronically inflamed, with or without coexistent mucous inflammation, it is enlarged, or, as Dr. Bennett has it, hypertrophied, at first not very hard, but if the inflammation continues, there is hardness; hence we have hypertrophy, induration, and enlargement. Hypertrophy is not the word for this condition of things; the part does not enlarge by an increase of existing tissue or a development of more of the same kind, but it is enlarged by an effusion of fibrin, which assumes an imperfect arrangement. It is increased in size in this way and also indurated. This kind of enlargement should be distinguished from the enlargement of congestion, a condition in which the uterus is injected with an unusual quantity of blood, and its substance distended by it. This is the case every month, but it becomes more permanent by the continuance of some point of irritation which keeps up an afflux of blood, and yet the irritation is so moderate as not to induce that stress of circulation necessary to an effusion in the tissues. We can, therefore, have chronic enlargement of the neck, and even the body of the uterus, without induration or actual structural changes. This is often the case where the

Fig. 2.

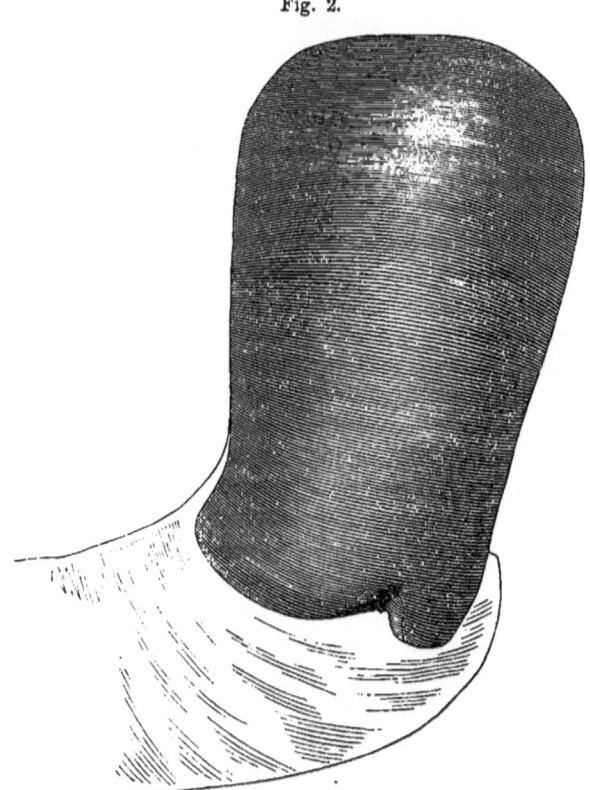

Induration and Enlargement from fibrinous deposit in submucous inflammation.

inflammation is confined to the mucous membrane. Enlargement is no evidence, therefore, of fibro-cellular inflammation; induration must be superadded to make the whole of the changes necessary to constitute a case of it. When, therefore, we meet with an enlarged and *indurated* uterus, or cervix, we may with safety conclude that it is suffering under chronic inflammation of the fibro-cellular tissue, with certain provisions that I shall have occasion to mention in future. When the uterus is hypertrophied, as in pregnancy, or in consequence of a growth or

Fig. 3.

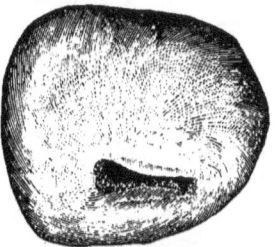

Appearance of Os and Cervix in enlargement from fibrous induration.

other substance which causes a development of tissue, the fibro-cellular structure is softer than natural.

Fig. 4.

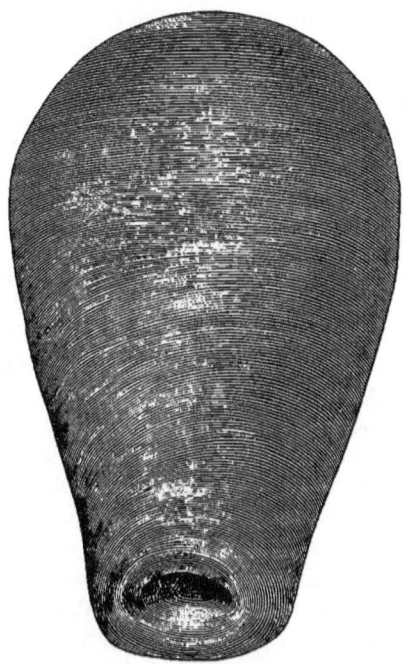

The appearance of the Uterus in a state of general engorgement without fibrinous infiltration.

Hypertrophy.—The hypertrophy from growth is often general, including the neck, body, and fundus; the enlargement from congestion is, most frequently, general, including the cavities; the enlargement accompanied with induration, and indicative of chronic inflammation, is apt to be partial; confined to the cervix, sometimes to one lip of the os uteri, or some part of the body near the neck. When the whole cervix is chronically inflamed, it enlarges in every direction; the thickness is increased from the size of the end of a man's thumb to half the size of his fist, or even larger than this, and it is hard and tender to the touch. The cervical canal is decreased in calibre in most instances, and somewhat lengthened. The induration is not always of the same intensity; its hardness is often very great, at other times but little more than natural. As the induration and enlargement may be quite partial, the shape, as well as size of the neck or portion of the body attacked, will seldom present its natural contour, and is frequently misshaped. The *proportions* of the different parts do not correspond in shape or size as they do in the healthy condition. Then we have in chronic inflammation of the different parts of the uterus, increase in size, hardness, and disproportion of corresponding parts, and hence alteration in shape, to which is almost always added tenderness upon pressure or touch, particularly with instruments.

Hardness with Atrophy.—Although these statements will be found to correspond with facts so frequently as to constitute the rule with regard to the subject, yet there are important exceptions. I have observed quite a number of instances, in which long-standing inflammation of the body of the uterus seemed to have brought about a shrunken condition of the organ. So that, notwithstanding the presence of all the symptoms, the uterus was very much diminished in size. It appears in such cases, also, to be indurated as well as decreased in its dimensions. It is barely possible these uteri were congenitally smaller than usual, and what seemed to be atrophy was natural. In many instances, I have had assurance that in the early part of married life there had been pregnancy and abortions. If this was true in these cases, they must have been pathological. It has been supposed that the fibro-cellular form of uterine inflammation always precedes inflammation of the mucous membrane for a greater or less length of time. This is certainly not always the case; for we meet with inflammation of the mucous membrane entirely unconnected with that of the submucous tissue, as a simple affection. They are, however, much more frequently combined than separated from each other.

Mucous Inflammation.—As a simple affection, that of inflammation of the mucous tissue is much the most frequent. Where they coexist, we have the increase of size, hardness, and irregularity of shape, indicating inflammation of the submucous substance combined with the evidence of mucous disease.

Seat of Mucous Inflammation.—The inflammation of the mucous membrane may extend to the whole of it, from the fundus through the cavities of the body and neck to the os, and then cover the whole of the vaginal portion of the uterus. This extent of inflammation is not very frequent, however, and when it occurs it almost immediately succeeds parturition or abortion, or is produced by gonorrhœal inflammation. I have seen it under these circumstances oftener than any other. It almost always causes a great deal of distress and suffering.

Cavity of the Cervix.—Probably the most common extent of inflammation is to the mucous membrane of the cavity of the cervix, and a portion or the whole of the membrane covering the intra-labial portion of the os. By far the greater number of instances that have come under observation in practice were inflammation of the membrane around the os and inside the cavity of the cervix. I fear that this statement represents a fact that has not been generally apprehended by practitioners. I am disposed to believe that too many physicians have failed of success in curing their cases, because they have not followed up the inflammation sufficiently above the os, in the cervix, being satisfied with curing that which was visible only, and, in consequence, leaving really the most important part of the affection untouched.

Cavity of the Body of the Uterus.—Sometimes the inflammation is limited to the cavity of the body, to the cavity of the cervix, or to the membrane in and external to the os uteri. Inflammation limited to the cavity of the body of the uterus is not common, but I am quite sure that I have met with it in some instances. Some of these had been treated for inflammation of the os and cervix, and cured of this, but the inflammation in the cavity of the body was left. Others had not had any treatment, as far as I could learn, for uterine disease. They had habitual leucorrhœal discharge of rusty-colored mucus, very much like the brickdust sputa of pneumonia; the os externum was very small, and the os internum uteri large, as was also the cavity of the body. One patient did not menstruate, and had not for a number of years, and although married, did not become a mother; the disease was caused by miscarriage in early life. She was thirty-four years of age.

Endocervicitis.—Endocervicitis alone, or inflammation limited or confined to the cavity of the cervix, is, on the other hand, an extremely common form of the disease. Not unfrequently this form of inflammation exists without any appearance of it in the os or external to it. When inflammation of the mucous membrane of the cavity of the cervix alone exists, it has certain effects upon the shape and other properties of the neck that are apt to attract our attention. Dr. Bennett describes the os as patent and the cavity of the neck enlarged, so as to admit the finger and permit the opening of it by a speculum to some extent, so that we may see the inside. Now, while this is very generally the case, it certainly is not always so. This open condition of the os and cervix is more frequently met with near the menstrual periods than at any other time, and is probably always owing to the congestion of the vascular tissue of the cervix and about the os.

Endocervicitis with Diminished Size.—I have undoubtedly seen many cases of this endocervicitis, in which neither the os nor cervical cavity were in the least enlarged, and others, in which the os uteri was contracted much below its natural size. The secretions of the mucous membrane are always modified; generally they are very much increased, and often changed in character. They may become purulent or sanguineous, owing to the grade of the inflammation and the degree of congestion. The inflammation situated external to the os on the end of the uterus, between the labia or their external surface, is very common, but it is not often limited to this part. It is almost always combined with endocervicitis.

Certain forms of these mucous inflammations are found more frequently in certain sorts of patients.

Endocervicitis in Virgins.—Virgin patients seldom have inflammation external to the os uteri; their disease is endocervicitis almost always; very rarely there is a little rim of inflammation around the os upon the end of the uterus.

Endocervicitis in Aged Women.—Again, in senile patients, women who have passed the climacteric period, and ceased to menstruate for some years, we find the inflammation in the cavity of the cervix. The os uteri in the aged is normally small, and simply looking at it will seldom convey a correct idea of the state of the cervical cavity, but the introduction of the probe in cases of endocervicitis will give rise to very great pain. The endocervicitis of old women is extremely difficult to manage, and is always protracted in duration.

External Inflammation combined with Internal in Childbearing Women.—In the married, childbearing women, we find the external inflammation combined with the internal uterine inflammation of the mucous membrane. They are the kind of patients in whom most frequently the enlargements, indurations, and fibro-cellular inflammations are observed. The form of disease in persons who have been married, but never been pregnant, partake to some extent of the character of both the virgin and the childbearing woman. They often have external, combined with internal, mucous inflammation, but not often fibro-cellular. Now, what I mean by these statements is, that these kinds of patients are likely to have the forms of disease which I have ascribed to them, but there certainly are exceptions to all of them.

CHAPTER VIII.

PROGRESS AND TERMINATIONS.

The intensity, terminations, and effects of inflammation upon the parts immediately implicated, of course will vary very greatly indeed. There can be no doubt that occasionally suppuration takes place in the fibro-cellular tissue of the uterus, especially the neck; but that such an occurrence is very rare is also true.

Progress in Submucous Tissue.—The inflammation of the submucous tissue seldom proceeds any further in pathological changes than an effusion of fibrin, and its more or less complete solidification. When once arrived at this stage, it is likely to continue indefinitely unless interrupted by some artificial or naturally intercurrent circumstance. The tendency of inflammation of this tissue is not to stop short of fibrinous effusion, and remain stationary for any length of time; it is either resolved before or soon attains to it. Whether inflammation commences in the deeper tissues, and affects the mucous membrane secondarily, is a subject which cannot be very often demonstrated. The probability is that this is occasionally the case; but what occurs more frequently, I think, is the transition of the inflammation from the mucous membrane to the submucous tissue, particularly in the neck and about the os. Hence it will be found that a case, as I have verified more than once, which this year presents only a tolerably bad form of mucous inflammation, without any tumefaction or hardness of the neck, in twelve months will present the tumefaction and induration characteristic of fibro-cellular inflammation.

Intensity of Mucous Inflammation.—In the mucous membrane the inflammation continues for an uncertain length of time without complicating the other tissues, and there is a very great difference in its intensity and effects. We often meet with inflammation of the mucous membrane of sufficiently mild grade to merely cause a slight increase in the color, heat, and ·secretion, without producing tissual changes. One thing which ought to be remembered, and I shall not apologize for the reiteration, is, that a permanent increase of secretion in a mucous membrane is, and should always be, regarded as an evidence of inflammation in it. Another not less important fact is that discolored mucus, either yellow, red, or otherwise, is not produced by a mucous membrane which retains its tissual integrity. Blood cannot get through the capillaries of a sound membrane, and pus is not produced by a mucous membrane while the epithelium retains its perfect integrity.

Progress of Mucous Inflammation.—After the inflammation has lasted for a time, if its intensity is increasing, the epithelium gives way more or less completely. The destruction, or rather the want of reproduction of epithelial scales, is generally observed in patches. At the point where the inflammation attains to the greatest intensity, the epithelium is not maintained. However small this point may be, the redness is increased; and if we look at it we see that the place is scarlet instead of a pale rose color, as when the epithelium is entire. Inasmuch as this is loss of substance, although slight, it is ulceration, or abrasion, the beginning of ulceration. As yet, the secretion is merely increased in quantity, or, at most, very slightly discolored with pus-globules, and rendered a little thinner by the exudation of a small amount of serum. The absence of epithelium is generally observed, where it occurs, in one continuous patch, of greater or less dimensions, and indicates not a very intense degree of inflammation. When this effect of inflammation is first observed, it is apt to be situated around the os uteri; but I have occasionally seen it over the whole intra-vaginal portion of the neck. The cases in which I have observed this state of extensive abrasion, were in persons who had passed the climacteric period of life, and they were the subjects of copious, watery, very irritating, and slightly yellow leucorrhœa; and upon examining them I was forcibly reminded of the chafed condition of the thighs in fleshy persons, so red and fiery was the appearance presented by it. They were obstinate, and it required great care in the use of remedies not to aggravate the inflammation.

Forms of Ulceration.—This epithelial denudation is the simplest and the most common form of ulceration met with in practice.

Of course, in this form the red portion is not depressed; it retains its level with the adjoining surface, and consequently the term ulceration is not considered applicable to it, by those who do not believe in uterine ulceration at all. After this description, as faithful a one as I can give, the reader will form his own judgment. I hope I may be allowed to consider it a breach of continuity of the mucous membrane, while anatomists persist in describing the epithelium as a part of that membrane. After the epithelium is lost for some time, there is a gradual increase in the size of the papillary structure of the membrane covering the neck of the uterus; and if the membrane is now examined, instead of the smooth redness there is something of a velvety or plushy appearance. The intensely red surface is covered by, or rather seems to be formed of, an infinite number of extremely minute projections, so closely apposed that there is hardly any space between them. Scarlet velvet is a very good representation of its appearance. The papillary projections do not seem larger than the minute silk fibres of velvet, as short and as thickly set. This surface is almost always covered with mucus and pus, in different proportions of admixture. There is always pus, however, when the complete absence of the epithelium is observed. Still, the evenness of the mucous surface is not disturbed. There is no excavation at least. If there is any change in this respect, the red patch is very slightly elevated above the surrounding surface. As the inflammation advances, the papillary development is greater, but also somewhat different; some of the papillæ increase faster than others, crowd upon the smaller ones, cause them to disappear, and usurp the space occupied by their oppressed neighbors. If the membrane is now examined, there will be seen, instead of the numberless minute, closely set papillæ, a greater or less number of larger ones, varying from the size of a small sewing-needle to a large pin's head, thickly studding the red surface. The redness now, as a general thing, is not so intensely scarlet. The ends of these papillæ, which rise from half a line to a whole line above the level of the surface upon which they stand, are darker red, inclining to lividity. The papillæ thus increase in size, and decrease in number, by strangling each other, until some of them attain the size of small shot, and look like warts. The larger they are, the greater is their lividity of color. As will be inferred, the diseased surface is more and more elevated and irregular, until very considerably raised above the surrounding level. In such cases, pus is generally poured out, in considerable quantities, from the spaces between the papillæ, and the whole surface is thickly covered with tenacious mucus, colored with pus, or with nothing but pus. Sometimes, however, such surface produces no pus or mucus, and seems preternaturally dry.

Complication of Mucous with Submucous Inflammation.—This sort of mucous inflammation is seldom observed without being accompanied by submucous inflammation as a complication. There is nearly always considerable enlargement and induration of the whole cervix where these greatly enlarged papillæ present themselves. In such cases as these, I think we may safely conclude that the inflammation commenced in the mucous membrane, and passed from it to the deeper structures. But there is another kind of enlarged, hardened neck, which with equal certainty begins in the fibro-cellular tissue, viz., when in connection with great hardness and enlargement, the surface deprived of its epithelium is extensive, and is uneven, or nodulations of moderate elevation, but greater extent of superficies than the papillæ, exist, reminding one of the rough surface of very coarse sacking or sea-grass carpeting.

Ulceration and Enlargement.—This kind of a surface is always seen upon a greatly enlarged cervix, which also is very much indurated. It is a very obstinate and very discouraging state of the disease, but will usually yield to sufficiently energetic and long-continued treatment. The boldness in the use of caustics necessary to the cure of such cases as these, requires strong nerves to institute and thoroughly execute. In the varieties I have here noticed, the surface is more or less elevated. But instead of papillary development after the destruction of the epithelium, the integrity of the mucous membrane is further invaded; the surface becomes somewhat depressed, with the edge of the red portion well defined; in short, ulceration, as it is usually understood, becomes quite evident. I should have stated before that in many cases, where the epithelium only is destroyed, the red patch shades off imperceptibly into the healthy rose color; in this last kind of ulceration the termination of the two is more abrupt.

Aphthous Inflammation.—Other sorts of ulceration occur less frequently on the neck of the uterus. Isolate, small ulcerations, several of them set upon a red surface, not unlike what we see upon the lips and inside of the cheek; also there are occasionally aphthous, or curdy spots, elevated somewhat, but soon degenerating into little yellow ulcers. I have, on one or two occasions, seen such ulcers in patients who were the subjects of nursing sore-mouth, and I always regard these minute isolate ulcers as the effect of constitutional disease; or they at least receive their peculiarity from the condition of the system, and indicate a general unhealthy state of the mucous membrane. It would hardly be proper for me to stop here to describe all the particular sorts of ulceration that occur; in addition to those resulting from inflammation, there are some which are the effect of specific diseases. The specific ulcers do not assume any peculiarity, nor are they particularly modified by their location upon the neck of the uterus. A chancre possesses its characteristic, when planted upon the neck of the uterus, as distinctly as when seated upon the glans penis. There is no difference between the peculiar, ragged, insensible, foul ulcer of scirrhus on the neck of the uterus and the mammary gland. The phagedenic ulcer of the uterus is the same as when observed to dissolve down so rapidly the tissues of organs elsewhere. A very little experience, care, and reflection, will save anybody from error of diagnosis or treatment, when the question of difference between common ulcer of inflammation and specific ulcer presents itself.

CHAPTER IX.

DIAGNOSIS.

FORTUNATELY for suffering woman, we may arrive at demonstrative knowledge of the extent, nature, and locality of diseases of the uterus; and, as a consequence, treat her diseases with the certainty which a positive diagnosis always insures. The evident advantages of a physical diagnosis will render it quite unnecessary for me to use any argument in favor of it, or to induce medical men to resort to it. A physical examination, however, of the genital apparatus of females, is quite a different matter from a physical examination of the chest, eye, or ear, or any other organ of the body; and hence the necessity of approaching and conducting it under conditions rendered imperative on account of the circumstances connected with it. The education and natural sense of modesty, so appropriate to female character, and which always command the respect of *gentlemen*, make such examinations disgusting and disagreeable above almost all others demanded by the necessities of woman's circumstances. With a view to this fact, it is our duty, by our conduct toward our patient, and the management of the examination, to divest it as nearly as possible of every disagreeable feature. Medical men generally I think are, as they should be, actuated by the above considerations, and I fear that they are often so influenced by their own sense of delicacy as too frequently to abstain from the enforcement of essential investigations. This is an error we should always bear in mind, and I think we shall less frequently regret a thorough, although somewhat indelicate examination, when dictated by an honest and intelligent conviction of its necessity, than a neglect of such examination from too great a deference to the mere shame of our patients. We should not be, in important cases, constrained to take things for granted that we are not sure exist. Our bearing to female patients should be deferential, candid, and modest. She should be convinced by our demeanor that everything we do and say is strictly necessary and relevant to her case, and has its foundation in our solicitude for her welfare alone. Nothing, therefore, should be said or done but what is called for and obviously proper. This sort of treatment from her medical adviser will always command the confidence and earnest co-operation of an intelligent female patient. There should be a full and explicit understanding between the physician and the patient, as to the necessity of a physical examination, in what it consists, and how it is to be conducted. The good sense of the practitioner will enable him to judge whether he should commit the detail of explanation to the husband, or some other appropriate second party, or whether he impart it directly to the patient; all the circumstances of the case will enable him to determine this matter without much difficulty. After the preliminaries are disposed of, I would insist upon conducting the examination without exposure. It is needless in ordinary uterine examinations, and should be permitted only when the disease is upon the external parts. One position and kind of preparation, so far as the patient is concerned, will suffice for most cases, whether we wish to make a manual or instrumental examination. There is no necessity for the patient to unclothe herself.

Position of Patient.—She should lie down on her back across a bed, so that the breech will be very near the edge; draw up her limbs by flexing the thighs and knees, and place her feet, separated about twelve inches, upon the side of the bed very near the nates. In this position a sheet should be thrown over her, so as to completely cover her person, and hang down several inches below her feet, over the edge of the bed. If we wish to use the speculum, or our eyes in any way, the bed should be placed immediately before a large window, in which the light is not obstructed by blinds or curtains. Thus placed, by kneeling down before her, we can have free use of both hands, a matter of much importance in delicate manipulations. Let the patient be very near the edge of the bed, lest by reaching too far, our examinations may be difficult, if not imperfect. When we wish to make a manual examination, we have need of nothing further than a little oil. Our objects in making a manual or digital examination are to ascertain the position, size, consistence, and sensitiveness of *all* the organs in the pelvis; the presence or absence of anything that does not belong there; and if anything unusual is there, what are its properties, connections, and nature. Upon making examinations for the first time, the whole of this investigation should always be attended to.

Some practitioners prefer a table to a bed, and sometimes have them made of peculiar shape and size suited to their fancy. In some respects a table is better than a bed, and where patients visit the office of the attendant may be very conveniently used. There are also chairs constructed for this purpose that have some advantages over either. What is known commonly as the invalid's chair will answer well for ordinary examinations. The patients may take their seat in it, and the proper position may be effected by pressing the chair backward until the patient is reclining. Where the patient is visited at her own house the bed will be the most available and answer most purposes.

When very great sensitiveness of the parts renders it impossible for the patient to remain quiet on account of the severe pain she experiences during the examination, it will be far better to subject her to the influence of ether or other anæsthetic. In this way thoroughness is secured where it would otherwise be impracticable. In some rare instances an anæsthetic will be indispensable.

Digital Examination.—The mode of examining the pelvis with the fingers is of the utmost importance. After oiling the fore and middle fingers the index should be very gently introduced, and the examination conducted as far as possible with it; then the two should be introduced, with which nearly all the cavity of the pelvis can be reached. The index finger will not reach as far, by one and a half to two inches, as the two together. As the finger is introduced, it naturally and easily comes in contact with the rectum, which may contain fæces, and consequently will appear as a round, full ridge along the middle line of the posterior wall of the pelvis, or a mere soft fibrous cord, hardly perceptible to the touch. The full rectum is generally a healthy one, as the fæces cannot remain long in a rectum rendered irritable by disease. By pressing upon the rectum with the finger, we may ascertain the presence of inflammation by the increased sensitiveness; the organ is absolutely insensible to moderate pressure when in a state of health. We should seek for internal hemorrhoids, which are small tumors in the bowels, or the induration and contraction indicative of stricture; and, in short, examine it as completely as possible in this way. Next, we should turn our finger forward, pass it up behind the symphysis pubis, and along the front wall of the vagina, and as well as practicable ascertain the condition of the bladder. It may contain a calculus, or other foreign substance, or, what is very much more common, be inflamed. In the first case the foreign body may be felt by the finger. The examination is more complete if the fingers of the left hand are used to press into the pelvis from just above the symphysis pubis. The substance can thus be grasped by the fingers of the opposing hands. With the fingers of one hand above the bladder, and the other in the vagina below it, we press it and thus ascertain its sensitiveness. With the two fingers of the right hand pressing up by the side of the uterus, between it and the walls of the sides of the pelvis, first on one side and then the other, while the fingers of the left hand press downward toward them from above, so as nearly as possible to meet them, the cavity may be pretty thoroughly explored, and any unnatural substance or uncommonly sensitive tissue be easily discovered. All these manipulations should be performed with the utmost gentleness, remembering that rudeness may deceive us as to the sensitiveness of organs, as well as give the patient unnecessary suffering. But while we are gentle, we should be as thorough as possible. The main object, however, for which we institute these examinations, is to ascertain the condition of the uterus with respect to position, size, shape, consistence, sensitiveness, &c. &c. Where is, or ought to be, the os uteri and cervix, and how shall we find them? In the virgin, the os uteri ought to be in the middle of the pelvis, upon or a little below the level of the arch of the symphysis pubis, and within easy reach of the index finger, two inches and a half from the entrance of the vagina. We may know when we feel the neck of the uterus by its consistence, shape, size, &c. It has more consistence than any part with which our finger comes in contact, as we push it backward into the vagina. In passing through the vaginal canal, the finger is impressed with a soft intestinal sensation, and can distinguish nothing but loose folds

Fig. 5.

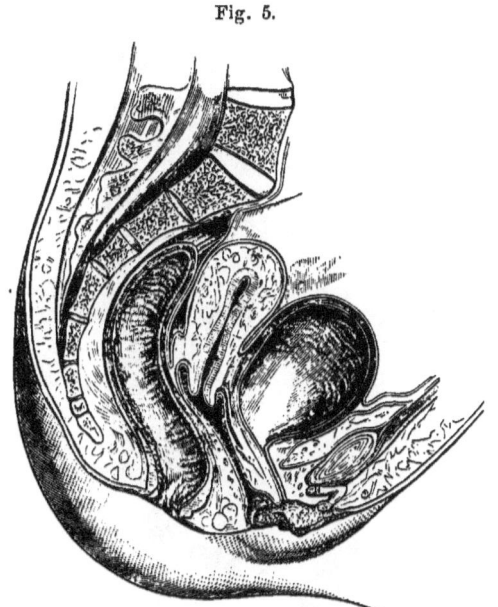

Natural Position of the Pelvic Organs.

that are dissipated and lost in the surrounding softness by the slightest pressure, until it comes to the neck of the uterus, when it may be felt having consistence enough to retain its shape under considerable pressure. If we push it upward, backward, or downward, it retains the same characteristics. The finger can be carried up the side, up before, or behind it as a projection, and surround it in every direction except above. This being unlike anything else in the vagina, will be easily recognized by an educated finger. The shape of the virgin cervix uteri is almost cylindrical, slightly compressed from before backward, and not far from three-quarters of an inch in diameter in every direction; it projects half an inch into the vagina, and the projecting or free end of it is apparently cut nearly square off, so as to present at its inferior face almost a flat surface, with a mere dimple in the centre corresponding with the os uteri. The cervix uteri of the childbearing woman

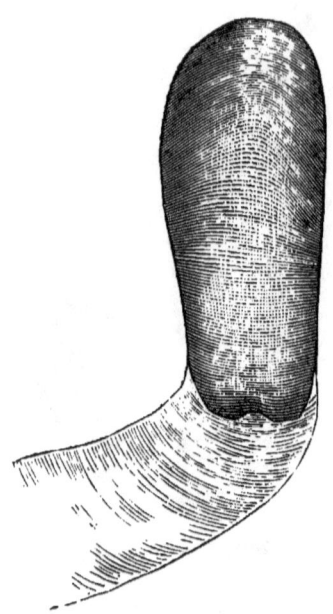

Virgin Uterus and Vagina.

is generally a little lower in the pelvis, and often slightly turned to one side, does not project so much into the vagina, is about an inch wide, or often a little more, from half an inch to three-quarters in its antero-posterior diameter, and instead of being truncated, seems formed of two distinct projections at its inferior extremity (the anterior and posterior labia of the os uteri). Between the labia or projections is a deep fissure, with its extremities directed to the sides, large enough to partially admit the extremity of the index finger. *Os tincæ* is applicable to this form of the os uteri, but in nowise is expressive as connected with the shape of the virgin os uteri; neither is it descriptive of the senile uterine mouth.

Fig. 7.

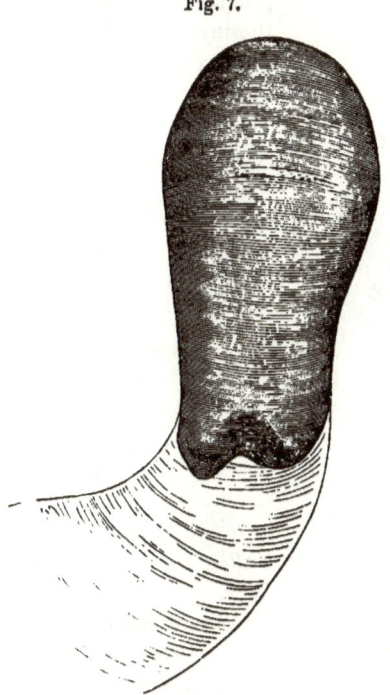

Uterus of a Childbearing Woman.

Os Uteri in the Aged.—The os uteri in the old is higher in the pelvis than in the virgin or multipara, does not project into the vagina, and feels more like a pit at the termination of the vagina. As women advance in age this description is more applicable than very soon after the cessation of the menstrual discharge. There is often a cord or frænum-like projection in the vaginal walls which is planted into the external surface of the anterior and posterior lips of the mouth of the uterus, and thus extend backward and forward to be lost in the anterior and posterior median line of the walls of the vagina. This frænum is more apparent, if not more developed, as women advance in age;

but I have known it so prominent as to be mistaken for the results of disease, even in the middle-aged. In one case an intelligent practitioner thought it an evidence of the injurious effect of strong caustics. The consistence of the virgin and multipara cervix uteri is the same. To the sense of touch it gives the idea (which is a correct one) of deep fibrous tissue, almost as hard as cartilage, covered over thickly with areolar tissue. Dr. Bennett compares it to the feel of the cartilage of the lower extremity of the nose. It seems to me not quite so dense, although nearly so. In health it is wholly insensible to pressure with the pulp of the finger, and it requires considerable force to produce pain with a plain round instrument. This fact should be borne in mind in our examinations, viz., *a healthy cervix uteri is not tender to the touch.*

Fig. 8.

Senile Uterus and Vagina.

Corpus Uteri.—We may examine the shape, size, and sensitiveness of the body of the uterus by pressing it down well into the pelvis with the left hand, while the fore and middle finger of the right presses upon it as high up as possible. When the uterus is healthy, the fundus cannot generally be felt above the symphysis, even by lifting it with our fingers, so that if it can be felt by both hands it may be considered enlarged.

Tender Uterus is an Inflamed Uterus.—I cannot refrain from emphasizing the fact that the uterus is insensible to the handling of an ordinary examination, and that a *tender uterus is a diseased uterus*—in fact, generally inflamed. What condition converts comparatively insensible organs elsewhere—the periosteum and cartilages, for instance—into highly sensitive ones?

A digital examination through the rectum will sometimes throw much light on the condition of the uterus. The index and middle fingers on the former alone, introduced their full

DIAGNOSIS.

length into the rectum, will reach high up the posterior surface of the uterus, and when retroverted may be extended entirely above the fundus and meet the point of the catheter directed backward and downward through the bladder. They may also survey the regions of the ovaria on either side, and discover disease or effusion in the folds of the broad ligaments as well as those organs, especially when the parts are pressed down well into the pelvis by the left hand from above the pubis.

Examination of the Urethral Canal.—If we have gained all the information we can from the use of the fingers, we may next use the probe, for the purpose of penetrating the cervical and uterine cavities. When, from the sense of touch, there is suspicion of inflammation of the urethra, the probe may be used with great propriety in examining this canal. There is almost always pain when the probe is introduced into the healthy urethra, but it is a peculiar smarting pain; if the urethra is inflamed, it is a sore pain; it feels as though the probe had touched a sore place; it is soreness. Dr. Simpson first recommended and practiced the use of the probe for the purpose of probing the uterus, and he has given to it a certain appropriate shape, size, and adjustment, which add very considerably to its usefulness and adaptability to this particular use. It may be found in almost any of the shops of our instrument makers, under the name of Simpson's uterine sound.

Object in Using the Probe.—The main objects in examinations with the probe in such cases as I have now under consideration are, to measure the size and length of the cervical and uterine cavities, the mobility and position of the uterus, and, if need be, the connection of that organ with pelvic growths. The instrument must be adapted to these purposes; in order to this it must be long enough, of the right size, and made of flexible metal.

Size and Length of Probe.—It should be ten or twelve inches long, with one end fixed to a flat handle; the probe end should be terminated with the ordinary

Fig 9.

Uterine Probe, or Sound.

probe-pointed enlargement about one-eighth of an inch in diameter. The wire behind the bulbous termination should be one line in diameter, round and smooth, and should gradually increase in size to the handle, where it ought to be about a quarter of an inch in diameter. The best material, I think, is copper, galvanized. I have not spoken of notches and other scale-marks upon it, because I like it better plain. Yet I see no objection to them as recommended by Dr. Simpson. It is always well to have two or three sizes of probes for special purposes, but the one I have here described is the one I should recommend to arrive at any deviation from the natural uterine measurement.

Mode of Using.—After oiling the instrument, and introducing the index finger of the right hand, and placing it upon the os uteri, the probe may be carried along the palmar surface of the

Fig. 10.

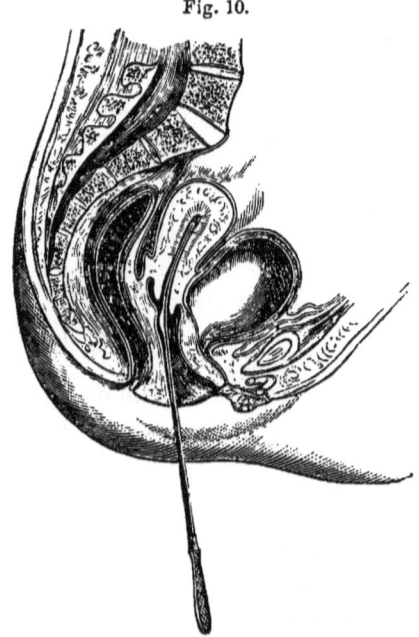

finger until the point arrives at the mouth of the uterus, when, by elevating the point, it may be carried forward into the cavity of the cervix. In order to insure its passage through the cavity of the cervix, into the cavity of the body, the probe must be bent to the same degree of curvature as a male catheter. Great gentleness must be observed in the use of this instrument, because it is an easy matter to do violence to the mucous membrane by a very little rudeness of management. After the probe has passed to the os internum, a sense of constriction is felt through the instrument, which feeling soon gives way, and the probe then goes to the fundus without further resistance.

Length of the Cervical and Uterine Cavities.—The cervical cavity in the virgin is about an inch and a quarter in depth, and the cavity of the body from a half to three-quarters of an inch; the former in the multipara is one and a half inches, and the latter an inch deep. In old age, they both are nearly or wholly obliterated. I do not often use the probe in this way for the examination of the uterus in cases of inflammation and ulceration, but have adopted the suggestion of Prof. Miller, of Louisville, and use it through the speculum, and shall consequently have more to say about it in connection with the use of that instrument. To expose the neck of the uterus so as to examine it by the sense of sight, it is necessary to have a speculum, and we ought to have a pair of long light dressing-forceps also; they will be very useful on several accounts.

It often happens, with the present means, that there is great difficulty in determining the thickness of the uterine walls, and even the presence of a small growth in the anterior or posterior parietes. For the purpose of enabling the inexperienced to arrive at what, in many instances, is valuable information in this respect, I have devised what may be called the uterometer, a cut of which is here given. It consists

Fig. 11.

The Uterometer.

in the adaptation of two uterine probes to each other, with handles and scale for measurement, in such way as that one may be introduced into the bladder, and the other into the rectum, and approximating them on the uterus as represented in Fig. 12, and then fixing the handles and scale in such manner as to make the measurement. When this is done, the instrument

Fig. 12.

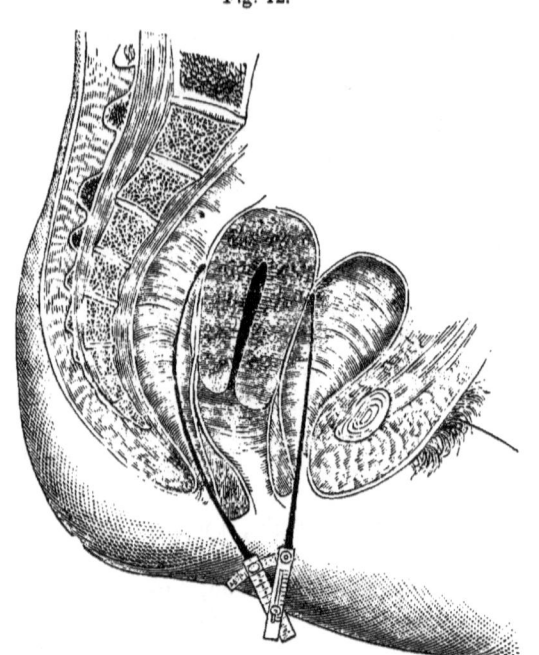

The method of applying the Uterometer for measuring the thickness of the uterus.

may be detached, withdrawn, and the exact thickness of the uterus is ascertained. If we wish to measure the posterior wall, one probe is introduced into the cavity of the uterus, and the other into the rectum, and the scale and handles adjusted, the measurement taken, and the instrument withdrawn. When the anterior wall is to be measured, one is introduced into the uterine cavity, and the other into the bladder. In this way, the length of the uterus and the thickness of the walls may be easily measured.

It is believed that the simplicity of this little instrument will enable us to be much more accurate in our estimate of the shape of the uterus, than any other means we can employ. The handles of the probe are adapted to each other by means of a slot, running from one end to the other, in one of the handles, while the other is of a size to fit into this slot closely and accurately. The scale is made movable, and may be easily adjusted after the probe portions of the instrument are in their proper place.

In cases of distortion of the cavity of the uterus, or when there is a tumor to measure, the probes will be bent in different directions, until they adapt themselves to the shape of the parts. In consequence of the necessity of variance in the curvature of the probes in making such measurements, the scale can serve only as an index to the relative position of the two probes, and cannot be relied on for the exact size of any growth or other cause of thickness of the walls. After having adjusted the scale, therefore, and observing the figures, we must withdraw the instrument and readjust by the scale, and then measure the distance between the points of the probes. This will give us the true measure. Sometimes, in fact, often, the instrument may be withdrawn with loosening it, which fact will facilitate the process very much.

In cases of retroversion or retroflexion, when we wish to diagnosticate this displacement from a small tumor, which they sometimes very closely simulate, one of the probes in the bladder, so curved as to follow downward and backward the anterior wall, the other in the uterine cavity, will clearly make out the difference. In like manner, only with reversed curves, and one probe in the rectum, the tumor may be diagnosticated to be present or absent. If I am not mistaken, this little instrument may be turned to valuable account in many obscure conditions of the organ.

Speculum.—Since the speculum has come into so general use, it has assumed a variety of shapes, and been composed of quite a number of different sorts of materials. For different purposes it is convenient, if not necessary, to be provided with different shapes, sizes, &c., of this instrument, but for ordinary use, the common cylindrical, or quadrivalve, are the best forms. My preference is for the compound called "German silver." If we use the cylinder, we ought to have three different sizes: one small, one large, and the other a medium size. With regard to the adaptation of the cylindrical instrument, the larger size we can use in the case the better, as it will the more completely ex-

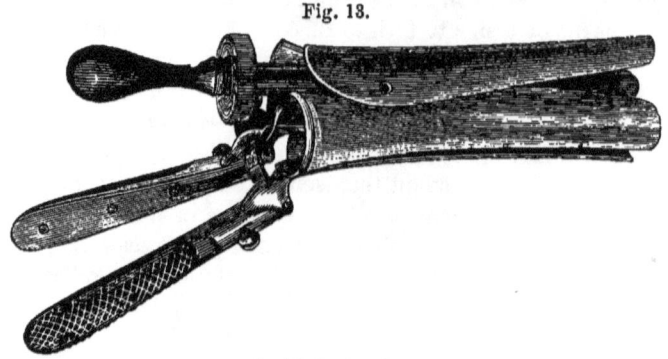

Fig. 13.

Quadrivalve Speculum.

pose the cervix. In selecting the cylindrical-shaped instrument, we should procure one with as great bevel at the internal end as we can find. There should also be always adapted to it a wooden director. The instrument will pass the external parts with less pain, and will not require the care to prevent it from injuring the vagina, than it will without the director.

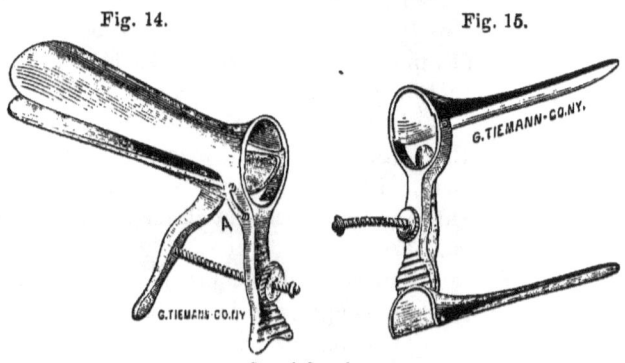

Fig. 14.   Fig. 15.

Storer's Speculum.

The common glass instruments, whether plain or covered, as recommended by Fergusson, I never use, excepting for the purpose of leeching. The instrument I prefer above all others for ordinary use, is the quadrivalve made according to Charriere's plan, or what is equally as good, perhaps, Tiemann's new quadrivalve. The former has a plug, or director; the latter is so ar-

Fig. 16.

Glass Speculum.

ranged at the ends of the blades that they close in together, and make the internal end of the instrument smaller, and render the director unnecessary. One quadrivalve will answer almost any case that occurs, if it be of medium size. It is constructed of four blades, that are caused to open or close by the use of a nut upon a screw. After introducing the instrument, the internal end may be increased greatly in size, without the external end being enlarged at all. It is only necessary to see it to perfectly understand its construction and the mode of managing it.

Position of Patient for Speculum.—To be prepared to use this instrument to the best advantage, our patient should be placed in the position I have heretofore described, viz., across the bed, before a large window, through which as much daylight should be freely admitted as possible. The better light the better view, and unless we have plenty, we cannot be certain of correct results in our examinations. The bed and patient should be so placed that the light may fall straight through the instrument, and full upon the parts at its internal extremity. We should also have some cotton-wool, sweet oil, and a couple of napkins, together with the dressing-forceps I have before spoken of.

Mode of Using the Speculum.—In commencing the examination, we should oil our speculum, and our middle and index fingers. Kneeling before the patient, we should introduce the index finger, and, if need be, the middle one also, to ascertain the position of the cervix uteri. This precaution will enable us to know in what direction, and how far, to introduce the speculum. After this preliminary examination, the forefinger and thumb of the left hand should be placed upon the edge of the labia, one upon each side, with which they should be gently separated;

holding the speculum in the right hand, somewhat like a pen, we may introduce it by the guidance of the thumb and finger placed as above. In introducing it, we should push it forward sufficiently to reach the cervix, and direct it upward, downward, or to one side, as we may have ascertained, by digital examination, to be the position of the os and cervix. The director may now be removed, when the os uteri will be seen at the open end of the tube.

Fig. 17.

How to find the Os Uteri.—If this is not the case, we may use our probe, and gently push the parts from one side to the other, turning the speculum in different directions, until it is found. If the neck is too large to enter the speculum, we may spread the blades still more, until it is brought into full view. Most frequently the parts are covered with some sort of secretion, and we should always, with cotton-wool or lint, remove all of it, so that the naked mucous membrane alone presents itself to our view. Without this precaution, we may overlook an obvious and extensive ulceration; for as the parts are covered over with this thick, opaque secretion, it either completely hides the parts from view, or much modifies their appearance. I have often met with cases where I have observed them attentively, for the purpose, if possible, of detecting ulcerations without this step, but failed, until the cotton was used, when extensive ulceration appeared. Indeed, I never think of coming to a conclusion of any kind by the use of the speculum, without this precautionary measure. By means of the sight, we can see the color, size, shape, and some other conditions of the parts, and the color, consistence, and derivation of the secretions. When the mucus, pus, or blood, comes from the mouth of the uterus, we can see it issuing from it. The shape and size of the neck and os of the uterus differ in different individuals, according as they have been impregnated or not.

Dr. J. Marion Sims pursues a different method from this in making examinations. He prefers a table to a bed. The patient is placed on the left side, the left arm under and behind her, the legs strongly flexed upon the thighs, and these again upon the abdomen, while the right knee is thrown forward, and over the left one on the table; this turns the patient over on the chest and partly on the abdomen. In this position his speculum is in-

Fig. 18.

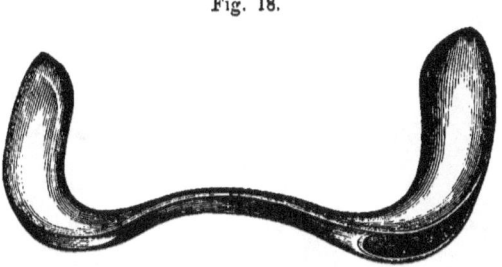

troduced by placing the forefinger of the right hand in the concavity of the extremity to be used, and the finger and instrument introduced together. When well inserted the perineum is drawn

Fig. 19.

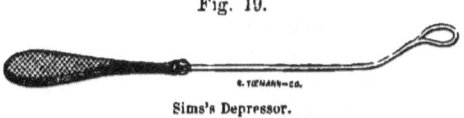

Sims's Depressor.

backward and the instrument is given to an assistant to retain in place. This will generally expose the cervix uteri completely, but if it does not, the depressor is placed upon the anterior wall, and this latter is pressed out of the way, as represented in Fig. 23. Great freedom of examination is thus obtained in most cases. Still, if the os uteri is not seen plainly, it is seized with a tenaculum and drawn toward the external orifice. Many practitioners prefer this method of exposing the organ for all ordinary purposes of inspection and application. Dr. Emmet has improved upon the speculum of Dr. Sims by constructing it in a fashion that renders it self-retaining, and thus does away with the necessity of having an assistant. Many other self-retaining instruments have been invented, that answer an admirable purpose, among which I mention, Dr. Pallen, of St. Louis, Dr. Nott, of New York, and Dr. Thomas. Of course it is necessary to have the patient so placed that the light will fall full into the dilated vagina

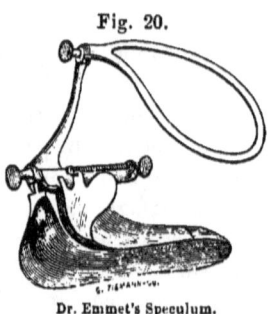

Fig. 20.

Dr. Emmet's Speculum.

Fig. 21.

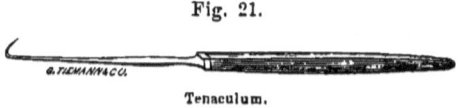

Tenaculum.

and on the cervix. Dr. Sims draws the cervix down when necessary, by means of a tenaculum; this often facilitates the ex-

Fig. 22.

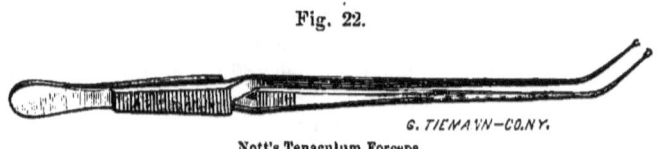

Nott's Tenaculum Forceps.

amination, and enables the practitioner to make applications or operations upon it with much certainty.

Appearance of the Os and Cervix in the Virgin.—The virgin uterus is small; the cervical end is nearly round, and terminates in a truncated extremity. Through the speculum, it does not present the appearance of labial projections, and the os is either a small slit about a quarter of an inch long, or a round opening into the middle of the truncated extremity. It is about large

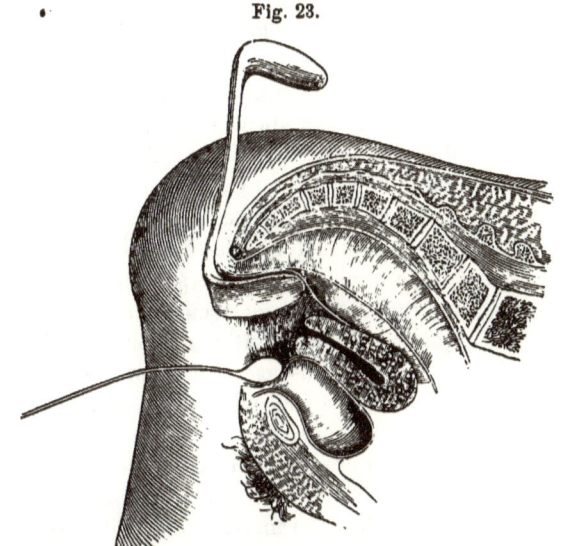

Fig. 23.

This figure represents the action of the instruments in Sims's method of examining the uterus.

enough to admit with facility the end of a female catheter, and the neck projects, in relief, from the bottom of the parts exposed by the speculum, something like half an inch.

Appearance of the Multiparous Uterus.—The appearance of the multiparous uterus is quite different from this; the cervix terminates in labial projections, which divide its extremities into an anterior and posterior half, and it does not project with so much prominence into the speculum. The os is represented by the cleft between these labial projections, and is large enough, in many instances, to admit the tip of the index finger.

Appearance in the Aged.—In the aged, the labial projections seem to have atrophied to obliteration, and the speculum shows a round opening in a funnel-shaped depression, surrounded by the walls of the vagina.

Exceptions to these Appearances.—Although the above is an accurate description of these appearances under the different circumstances, there are many natural deviations from it.

Color.—The color of the mucous membrane covering the cervix, and entering the os uteri, may be compared to that of the inside of the lips of the mouth, a pale rose red.

Appearance of Secretion.—The parts are merely lubricated, not smeared or inundated, with mucus. There is just enough of this secretion to keep the membrane moist, but not enough to hide the surface from view. I speak now of the cervix uteri.

Indication of Mucus in Abundance.—An abundance of mucus must be regarded as an evidence of excitement; its constant and persistent abundance as an evidence of disease. "Remember that in spite of their name, it is not the business of mucous membranes to secrete mucus; the more perfect their condition, the more favorable the surrounding circumstances, the less they do so. . . . The greater the diminution of their life, the greater the secretion." The more disease, the greater the secretion, until their integrity is destroyed, when the secretion becomes modified. The source whence this mucus is derived will show the point of disease; if it comes from the os uteri, the disease is in the cavity of the cervix, or body of the uterus.

Indication from Pus.—These remarks apply with greater force to the production of pus by the vagina or cervix. Pus or purulent mucus, indicates disease; and when we find muco-pus, or clear pus, in the end of the speculum, it would be preposterous to conclude that there was no disease there, merely weakness of the parts. It is extremely doubtful whether pus can be produced by mucous membrane, without destruction of the epithelium, at least. Temporary congestion often increases the amount of mucus to be found in the vagina, but no pus. The color of the mucous membrane, in cases of congestion, is a livid or a dark purple red, instead of the scarlet of abrasive inflammation.

Probe and Speculum conjointly.—When the neck of the uterus is exposed in the speculum, it will often be profitable to use the probe. If proper attention is paid to appearances under the use of the probe much information may be gained. When the mucous membrane of the cavity of the cervix or body is inflamed, it is generally much more fragile than natural, so that it bleeds upon slight contact with the end of the probe. In cases where the inflammation extends to the cavity of the uterus, the probe passes the os internum without obstruction, and passes farther up than natural from the increased size of the cavity.

Characteristic Signs of Inflammation.—The signs of inflammation of the submucous tissue or substance of the neck of the uterus are, increase of size, tenderness, and generally hardness; of the mucous membrane, increased color and secretion; of ulceration, still more intense redness, purulent discharge, tenderness, and generally enlargement. The former conditions may be ascertained by the touch, the latter by the sight, and when they are mingled, by both combined. It may be superfluous to pursue the diagnostic description of these cases further; but as I believe that a great many members of the profession do not sufficiently appreciate the importance of some of the appearances and conditions I have described, and as I am thoroughly convinced of their significance, I am determined, at the risk of reiteration, to place these diseased appearances and conditions in a more prominent light. Open external ulceration of the uterine cervix, after the parts are well exposed, *and cleared of mucus and pus* by wiping, cannot be well mistaken, or overlooked; and the only thing I shall insist upon here is that the practitioner must not be led to believe the case one of no importance because the ulceration is not very extensive. This raw scarlet surface is always indicative of mischief; and we should expect any amount of suffering from even small patches of it. There are cases where the appearances are not so obvious; where in fact all the parts exposed by the speculum and within reach of our vision have a natural appearance. No redness, rawness, or other discoloration can be detected on the neck, in the mouth of the uterus, nor on the vaginal surfaces; they are quite healthy in appearance and reality.

Diagnosis of Endocervicitis.—But there is an obvious and in many instances a copious secretion of tenacious mucus flowing from and lying in the os uteri; wipe this away and all looks right. This is a case of endocervicitis. In some instances this mucus is colored with streaks of yellow by the presence of pus, or it is wholly yellow; here there is loss of integrity in the epithelium of the cervical cavity. The mucous membrane in the cervical cavity is ulcerated. If we remember that the mucous membrane secretes only enough mucus for lubricating purposes, in the natural condition, we can arrive at no other conclusion than that the membrane is in a state of hyper-excitement when its secretion is abundant, or altered, or both. When we see mucus in even small, yet perceptible quantities, issuing from the anus, what is the inference? If this is abundant, persistent, and colored yellow, however healthy the anus might appear externally, we could not believe that the rectum was in a healthy condition. Why not then positively determine that the mucous membrane is inflamed, which floods the os uteri with mucus or pus, or with both? If we introduce the probe into the cavity of the cervix thus abundantly secreting, the patient will nearly always complain that we touch a "sore place; a tender spot;" that it hurts her in her back, &c., &c. And very often blood will immediately follow the withdrawal of the instrument. This, however, is not invariably the case. Another diagnostic evidence of endocervicitis, is the increase of the pain ordinarily experienced by the patient when the probe or nitrate of silver is introduced.

Diagnostic Effect of Caustic Applications.—There is not a new pain produced, but the old pain is aggravated, and the quality of the symptoms are the same while the number is increased. If the pain in the sacrum has been the one mostly complained of, the introduction of the caustic makes the back ache worse; if the pain in the iliac region has caused most suffering, it will be aggravated. The hyper-secretion, or perverted secretion of the mucous membrane, must then be regarded as an indication of disease of that membrane. If we have these facts fixed in our mind, and if we act upon them, we may discover and cure disease that would otherwise escape our attention, and thwart our skill. But there is another obvious and common-sense sign of inflammation which has not been applied in our investigations of diseases of the uterus, viz., tenderness. Tenderness or sensitiveness to the touch anywhere else, leads us to suspect inflammation, but in the uterus it is unaccountably set down as indicating an irritable uterus and not an inflamed one.

Diagnosis of Submucous Inflammation.—I think when I touch the uterus with the finger or an instrument, and the patient shrinks from the contact and says, "she is sore," or "it is sore," that there is inflammation there. Tenderness is not an evidence of mucous inflammation, but of submucous or fibrous inflammation of the uterus. And it is a matter of importance to determine the presence or absence of submucous inflammation either as an independent affection or complication of inflammation of the mucous membrane. It is a great error to confine our attention to the abrasions or ulcerations of the mucous membrane; and to believe we must see those abrasions or ulcerations before we can admit the presence of inflammation.

Complication of Mucous with Submucous Inflammation.—We should not shut our understanding to the fact that the uterus should be examined by the same diagnostic rules that govern our investigations of disease in other organs. Some authors tell us that ulceration results from inflammation of the submucous tissue; and others that the inflammation begins in the mucous membrane. However this may be, I am sure that inflammation sometimes exists in both these tissues at the same time. In this case we shall have tenderness and hyper-secretion. At other times there is submucous without mucous inflammation; then we shall have tenderness without hyper-secretion. Again, we may have mucous without submucous inflammation, when hyper-secretion without tenderness will indicate it. These remarks will fix the importance of these two symptoms as indicating the seat of the disease.

Size of the Uterus ordinarily increased—Exceptions.—The size of the organ would seem to be a good indication of the presence or absence of inflammation; but this may vary very much under what would appear to be the same form of disease. In endocervicitis it is usual to find the cervical canal increased in calibre; but this is certainly not always the case, as I have met with unmistakable instances in which this cavity was decreased in size and the os uteri almost closed; it was so small as to admit only a very small probe. Where there is mucous inflammation of the cervix extending toward the cavity of the body, and more particularly where the disease extends into the cavity of the body, the whole organ is likely to be enlarged. So much enlargement sometimes takes place that the fundus may be felt considerably above the pubis. Neither is this always the case, however; often there is no enlargement. The hypertrophy, or general enlargement of the organ, is more frequently indicative of mucous than submucous or fibrous inflammation.

Atrophy as the Result of Inflammation.—In fact, I think that long-continued inflammation of the substance of the body and cervix often brings about atrophy or shrinking of the uterus. Permanent increase of size or hardness of the cervix must be the results of submucous inflammation, and generally coexist with it.

Examine for Complications.—Some, if not all, of the pelvic or local symptoms, or such as much resemble them, may be produced and perpetuated without inflammation of the uterus; hence it is necessary to examine the case with reference to this fact. We shall also occasionally find that, notwithstanding the complete cure of actually existing uterine inflammation, the local symptoms, in a modified form, still continue. These circumstances will be found to depend upon the independent or coexistent presence of some of the complications I have described.

Cystitis, &c., as a Complication.—Chronic cystitis, rectitis, prolapse of the rectum, piles, urethritis, cellulitis, &c., &c., are among the most common causes of these symptoms. It is only necessary to mention these facts to enable the intelligent practitioner to explain anomalous cases that occasionally occur. There can be no doubt, I think, that holding the uterus to a rigid accountability for all the pelvic symptoms enumerated as the ordinary result of its diseases, has caused a good deal of confusion, and has enabled certain writers triumphantly to assure their adherents that in a number of cases the symptoms were present, but the ulceration was absent! A number of organs commanding extensive sympathies, sensitive under inflammation, crowded together in so small a space as the pelvis, supplied to a great extent with branches from the same nerves and arteries, must all be more or less congested, inflamed, and pained together; and nothing but an intelligent and deliberate physical examination can make out the difference in their relative suffering, or certainly ascertain which of them is affected when one alone is diseased.

Almost the only disease with which chronic inflammation and ulceration of the cervix uteri are likely to be confounded, is cancer in some of its stages. The many well-marked symptoms and physical conditions which accompany this last disease are now, however, so well understood and so thoroughly described, that the novice need not be embarrassed in his diagnosis of it.

I find in Becquerel's "Traité Clinique des Maladies de Uterus," pp. 320–323, vol. i, so complete and faithful a diagnostic summary between cancer and the different conditions of chronic inflammation of the cervix, that I have translated and given its substance for the concluding portion of this chapter. It is subjoined:

Cancer in the Scirrhous Condition.	Inflammation with Ulceration.
Cervix hard, unequal; nodulated, os not always open, sometimes wrinkled or furrowed.	Neck less hard, developed regularly in one of the lips, os always open.
Scirrhous of the neck often implicates the vagina.	The induration of the neck never extends to the vagina. Mobility of uterus complete.
Hereditary influence is often traceable.	No hereditary influence.
Touch is painless.	Touch painful.
Discharge sometimes absent; in certain cases very abundant, and consisting, for the most part, of albuminous serum.	Discharge constant, and characterized by the presence of transparent mucus, muco-pus, or purulent mucus.
Menstruation increased, being neither more nor less painful, and passing often into the state of real hemorrhage.	Menstruation more painful, often retarded, almost always scanty.
Absence of special anæmia when the vagina and body of the uterus are involved. Cancerous cachexia.	Special anæmia as above described.
Progress continuous and without cessation.	Often stationary for a long time.
The pain in cancer is very sharp, intense, and lancinating, and not influenced by locomotion or movements of any kind.	Pains less severe, more dull, and perceptibly influenced by walking and other sorts of motion.
Ulcerated State.	*Chronic Inflammation and Softening.*
Developed at the critical period of life generally.	Occurs earlier in life almost always.
Preceded and accompanied by hemorrhages.	Not preceded by hemorrhage.
Severe, sharp, lancinating pain.	Pain dull and profound.
Development essentially in sharp irregularities and nodosities.	Enlargement regular and rounded, or regularly lobulated.

Ulcerated State.	*Chronic Inflammation and Softening.*
Adhesions to other organs soon as ulceration is formed; immobility of the uterus.	Complete absence of adhesions to other organs. Entire mobility of the neck and body of the uterus.
The surface only slightly soft, subjacent tissue scirrhous.	Tissue of the cervix not hard, and easily destroyed.
Ulceration deep, unequal, essentially irregular, with thick, elevated, and hard edges.	When ulcerations exist, less deep, with tumefied edges.
Always granulations.	Granulation often accompanies the other lesions.
Discharges extremely abundant, consisting of purulent and often sanguineous serum; nauseous and often fetid odor.	Discharges less abundant, consisting of muco-pus alone, or accompanied with a little blood, without odor.
Great hemorrhage from time to time, not necessarily at menstrual period.	Always hemorrhage, but often a mere prolongation of the menstrual discharge.

Cancerous Ulceration.	*Simple Ulceration.*
Developed upon an hypertrophied and scirrhous surface.	Ulceration often on a healthy tissue, or presenting the soft or hard varieties of inflammatory injection.
Ulceration deep, vast, unequal, grayish surface with thick edges, and easily bleeding.	Ulceration more superficial, the edges less developed, and more regular at the bottom, not always easily made to bleed.
Ulcerated surface hard, presenting numerous lobes and tubercles, with nodosities and great hardness.	Nothing of the sort in chronic inflammation and ulceration.
Often great loss of substance.	Ulceration is not always accompanied with loss of substance.
Cervix and corpus uteri immovable, on account of adhesions.	Neck and body always movable.
Discharges sanious, fetid, sanguinolent, and of an insupportable and characteristic odor.	Discharge of muco-pus or purulent mucus; always less abundant.
Cancerous cachexia always present.	Special anæmia.

CHAPTER X.

GENERAL TREATMENT.

General Treatment.—I am sensible of the great difficulty of properly estimating the value of any given remedy or plan of treatment for the cure of disease. Nature does very much sometimes to aid imperfect means, and even to effect a cure under improper treatment; while at other times the circumstances inseparable from a case thoroughly thwart the best-directed efforts, and very often we record cures and attribute great efficacy to our plan of management, when the favorable termination is due alone, and perhaps in spite of us, to the natural conservative energy of the system or the parts concerned. It is often a mistake, therefore, to be too sanguine in our expectations even with the use of a favorite course of treatment, or to depreciate everything which has not fulfilled our hopes. We should patiently, honestly, thoroughly, and judiciously, try every means within our knowledge for the benefit of our patient, let him labor under whatever disease he may. The reader is doubtless perfectly aware of the very great differences of opinion in the profession as to the treatment most beneficial in inflammation of the cervix uteri and its accompanying ailments. In alluding to these many and diverse opinions, I must record my conviction of the honesty with which they are maintained by the principal disputants of the present day, and must exhort the junior members of the profession to cautious and thorough research on th subject. There must be a right and a wrong side to every disputed question; and, as a general thing, extremists are wrong. Remembering this general truth, we cannot always be kept in doubt by the facts in the case, if, without prejudice or party bias of any kind, we earnestly set to work to learn.

Spontaneous Cures.—Are there any spontaneous cures in these cases? I think there are, and I propose inquiring into the method adopted by nature, and take it as a guide to some extent, at least, for the plan of artificial treatment. Change of circumstances frequently makes robust persons of invalids. This change is generally from irregular improper habits of living to such as are regular and appropriate; from the highest state of luxury and ease to one of need, or at least economy and industry, in which the patient must exercise her mind and muscles, to a proper degree. The healthy tone of the stomach, muscles, and brain, thus brought about, decreases the susceptibility to slight suffering, enables the patient apparently entirely to recover from disease, and bear small ills without complaint. I need not specify the various circumstances and conditions of life which improve the tone and elevate the functional activity of the whole organism; they are numerous, and will suggest themselves to the reader. How many journeys are taken, how much time spent at watering-places and places of amusement for this purpose? And often they answer the purpose, and the patient is restored to health.

Change of General Circumstances only Temporary in their Effect.—This improvement in cases of disease of the uterus is brought about rather by diminishing the nervous susceptibility to the wearing influence and pain of the local disease, and by fortifying the system against its advance by establishing excellent general health, than by actual cure of the local inflammation. As a consequence we find a return to the former mode of living, habits, and circumstances, reproduces, more or less rapidly, the same train of general symptoms, and makes it necessary to resort to a repetition of the journey, or whatever other means were previously successful for their removal. This is only an apparent, not a real cure, and I hope I will be excused for saying that such is the kind of cures which always result from an exclusive general treatment. Tonics, laxatives, and alteratives, put the general condition of the patient on a better footing, and the patient suffers less from her local disease, and even considers herself well; but suspend the general roborant appliances, and the patient again sinks into her former state of valetudinarianism. I have often witnessed these changes as the effect of accidental mutation in the condition of the patient, intentional changes of place and circumstances, or well-advised general treatment.

Supervention of Acute Inflammation.—There is, however, another method resorted to by nature, and which sometimes results in permanent and complete cure. Chronic inflammation has very little tendency to spontaneous subsidence; its duration is at least indefinite. Situated in the neck of the uterus this is particularly the case. Acute inflammation, however, on the contrary, has a strong tendency to terminate in resolution, to subside and leave the parts in a healthy condition. And, in cases of chronic inflammation in any of the organs, the supervention of the acute form proves sometimes salutary. It absorbs the whole chronic action and takes its place in the tissues; and as it subsides, the diseased organ is left in a healthy condition. We have an opportunity of seeing this process of usurpation, displacement, or whatever else it may be termed, in diseases of the eye, and witnessing the salutary sequence.

Acute Inflammation after Parturition or Abortion sometimes works a Cure.—Some of the functions of the uterus when naturally performed are followed by acute inflammation in the neck of the uterus. I allude particularly to parturition; and while these inflammations sometimes linger and become themselves chronic, they generally, under favorable circumstances, subside kindly, and where the cervix had previously been affected by chronic inflammation, sometimes favorably modify if not entirely cure it. I think that very few cases of parturition occur that do not cause sufficient violence to the cervix and os uteri, to be followed by a greater or less degree of acute inflammation. A great many are certainly thus followed by inflammation. The acute inflammation resulting from abortions occasionally have the same effect. Instances have occurred in the hands of most experienced practitioners, where the uterine health of a primipara has been benefited by pregnancy and the processes of parturition.

Principles of Local Treatment.—The local treatment of these inflammations is founded on the same principle of these natural cures. In the case of obstinate inflammation of the eye, we often resort to strong stimulants to modify a chronic inflammation, *i. e.*, turn it into moderately acute one, which, usurping the place of the chronic, causes it to subside and leave the organ sound. And we know how successful it often is. So with the local treatment of inflammation of the cervix uteri. We awaken an acute inflammation in the tissues occupied by the chronic; and, as the former subsides, the disease is favorably modified, if not entirely cured. This is a radical cure, where a sufficiently strong impression is produced either by the natural or artificial process.

Physicians array themselves in two divisions in the treatment of uterine diseases. One division comprises those who consider the local disease as unimportant effects of the bad condition of the general health, who pay particular attention to the general condition of the patient, and who give but little, if any, local treatment. While the other division relies upon local treatment for the cure, and the general merely as accessory. Those who look upon the local as the essential treatment, are also somewhat divided as to the kind of treatment proper. One of these subdivisions thinks that if the uterus can be placed and sustained in its proper relative position to the other organs that the inflammation will spontaneously subside; while the other party believes in the use of strong stimulants and caustics applied directly to the diseased parts. I shall not at present pay much attention to the plan of mechanical support, leaving it for a future chapter, but will proceed to give the general and local treatment, which can be relied upon with most confidence for the relief of patients affected with inflammation of the cervix uteri, and I shall first give the general treatment.

Posture, Exercise, and Repose.—The young practitioner will soon learn that posture and exercise are important considerations in the general treatment, and he will be taught by most writers that the reclining posture and strict quietude must almost universally be observed. Walking generally causes an increase of pain, and, it is natural to suppose, an increase of inflammation; so that exercise on foot or in the erect position is regarded as injurious. On the other hand, confinement to the recumbent posture and the observance of strict quietude is very hard upon the general health; the patient becomes more nervous, and all her functions are performed in an irregular and imperfect manner. As a consequence, in very many instances, the symptoms are much aggravated. In the great majority of these cases, therefore, I think the patients are injured by confinement and recumbency. It would neither be scientific, sensible, nor successful, however, to lay down any absolute rule in respect to exercise and quietude. I think we may arrive at pretty accurate conclusions as to the sort of cases and the conditions under which each should be observed. More than ordinary acuteness of the symptoms, indicating a high degree of inflammation, occurring in the beginning and continuing throughout, or arising during the progress of a case as the effect of temporary causes, will make rest indispensable to the removal of them. Hemorrhage at the time of menstruation or between the menstrual periods, is also a reason for strict quiet. Where neither of these conditions are presented, I think the patient will be much benefited by judiciously directed exercise. I feel like insisting upon the enforcement of outdoor exercise as the rule in these cases; for I have often had an opportunity of contrasting, in the same cases, the influence of quiet and exercise upon the recovery of patients of delicate nervous constitutions. One patient who had been unable to sit up for even a short part of the day for several months, on account of the pain in the hips, dragging in the loins, and great nervous prostration, was sent to a water-cure, and in three months she returned home capable of walking several miles a day, and enjoyed comparatively robust health. In a few weeks after returning to a home in which she enjoyed the luxuries and ease so desired by all who prize good living, she became "miserable," and was obliged to abandon her exercise entirely. It is encouraging to state, that in less than six months of proper local treatment, she was permanently cured. This is but a type of many similar cases that have been benefited by the enforcement of exercise and other items of proper living, but, I must also add, not cured. It has been my constant aim for many years to induce patients of this kind to take as much exercise as they can bear. Under the mistaken notion that any local pain indicates an aggravation of their disease, and that to exercise when it gives them pain, even to a moderate amount, is a great evil, they confine themselves to their room, and even their bed, to the forfeiture of that healthy tone and energy of the nervous system which shield them from the intolerable and inexpressible *ennui*, melancholy, and irritability, which are so characteristic of bedridden women.

Pain and weariness, that subside after a few hours' rest, should not be regarded; it is only in those cases in which the pain and weariness increase at every effort at exertion that exercise should be abandoned, and then we should insist upon its being resumed again as soon as sufficient advance in the cure has been made to justify another attempt. We should not tire, during the whole treatment, in our endeavors to institute a system of regular and gradually increasing exercise, on account of the consideration that it is indispensable to the enjoyment of useful and comfortable health. Selection of the kind of exercise will depend, of course, upon the condition of the patient in respect to pecuniary matters as well as the state of her disease. Fortunately, the best kind is such as is within the reach of every kind of patients, not excepting those who are under the necessity of earning a living. The capacities and demands of our nature are formed to answer the curse pronounced against Adam. We not only earn our bread by the sweat of our brow, but the labor necessary to procure the bread brings almost all the conditions that insure health and happiness. It is, in fact, a great evil of the present state of society, that our ladies cannot find in useful employment that healthy tonic exercise for the body and mind which they need, and that such exercise and employment are allowable and acceptable only in amusement. There is almost no variety in mental and corporeal exercise required by the highest social amusements, and it is only when we descend to the primitive sports that our demands in this respect are met. It is undignified in ladies to fish, hunt in the woods, or engage in muscular feats. They must for muscular exercise engage in the measured sameness of the quadrille, or the giddy whirl and violence of the waltz, or cramp their limbs to the steady routine of a system of calisthenics. What are all these, for variety and adaptedness to their wants, compared to the washing, ironing, sweeping, milking, churning, spinning, weaving, cooking, walking, running, of household engagements: the stimulus of need; thinking of all these things; timing them; proportioning them; calculating, economizing, nursing, doctoring, advising, correcting, teaching, and conducting little minds and bodies through the physical, moral, and intellectual discipline which capacitates, unfolds, and imbues them with what is good and useful?

Woman's duties, taking them altogether, when well and appropriately performed, will do more than all the amusements that can be invented to keep woman well and healthy in every particular. In fact, it is only woman thus employed that can enjoy amusements. To the woman that constantly seeks after amusements, these very amusements become an irksome and toilsome business; they have a disagreeable sameness, and do not divert her; they simply vitiate her tastes. We all want variety, and constant employment, with a *sense of usefulness* attached to it. With this view of the usefulness of mental and bodily labor, I encourage my patients to engage in their domestic duties and labor, gauging the amount of labor by their capacity of endurance. Attention to the homes of wealthy women, as society is now constituted, requires a great deal of anxiety and mental exercise. Without a proper variety of muscular exercise, the woman, in attending to the duties connected with it, becomes more nervous; but the home of the poor industrious citizen or farmer gives enough and a healthy variety of both muscular and mental exercise to promote health and happiness. Should there be such objection in any shape as to make this course impracticable or improper, it is an interesting question to decide what sort of physical exercise is most desirable and beneficial. I am decidedly in favor of exercise on foot, outdoor, as one of the very best kind, far preferable to carriage or horseback riding. The carriage riding is not sufficient exercise for the most of such patients, and yet those who are most debilitated, and utterly unable to walk, may be much benefited by riding in an open carriage until they become vigorous enough to walk, when it should be abandoned. Convalescent patients may ride on horseback if they can have an easy-going animal; but this sort of motion is too violent; there is too much jolting for such cases until nearly or quiet cured of the local trouble. We ought to induce our patient to walk more each day than the previous one, if possible, until she has plenty of exercise.

Diet.—Unless during acute suffering, or on account of dyspepsia in some shape, a good substantial or nutritious diet should be allowed; and sometimes we may, with propriety, allow stimulating drinks; but as an ordinary thing, these last should be dispensed with entirely.

Sexual Intercourse.—Young physicians have often asked me whether sexual intercourse is injurious during the time of treatment, and whether it should be permitted? I have no hesitation in insisting upon entire abstinence from this act. The recovery of our patient will be more rapid, certain, and complete, when this is observed; and I believe that failures are the result of carelessness in this respect. It is very common for our patients to enjoy more comfort when absent from their husbands; and come home from a journey, as they think entirely cured, to be assured of the contrary, by the first effort at coition, and become miserable with pain, nervousness, &c., in a short time, on account of indulging in this conjugal act. I desire, therefore, to be explicit in warning my young friends in the profession, not to omit the interdiction of sexual intercourse, however delicate the task. A private interview should be sought with the husband for that special purpose.

Main Objects of General Treatment.—The main object to be gained by general treatment is to palliate the general condition of the system, to aid the local in effecting the cure, and to remove, when practicable, the effects left after a cure of the local disease. A *cure* of local chronic disease, by general treatment alone, is hardly to be expected; although, in some instances, it may be indispensable to such result. When general treatment is used as a palliative or adjunct in local diseases, it is directed to the relief of general symptoms attendant upon them. It will be impossible for me to notice the treatment necessary in all the symptoms which attend and add to the distress of our patients in uterine diseases; but there are certain prominent and troublesome ones, on which I cannot with propriety omit to dwell. I do so the more readily, from the embarrassment which I know, from experience, fills the mind of the inexperienced, as to the proper value to place upon general treatment, and the course to be pursued.

General Symptoms requiring Special Attention.—The symptoms, the treatment of which I propose to speak of in detail, are, 1st, general nervous prostration; 2dly, nervous excitability, exaltation of nervous excitement; 3dly, anæmia; 4thly, general plethora; 5thly, local plethora; 6thly, constipation; 7thly, indigestion. These are generally more or less complicated with each other, and sometimes several of them coexist; but, ordinarily, some one assumes the most prominence, and occasions most distress, and consequently requires more of our attention than the others.

Nervous Prostration.—There is often great nervous prostration, and a sense of weakness, when, so far as we can judge, hæmatosis and nutrition are usually well performed. What is the cause of this depression must be sought out in each case, as there is no uniformity in the functional deviations. Very frequently there is a deficiency of menstrual discharge, the scantiness being very obvious; at other times it is too copious. We should inquire into the functions of all the important organs, and correct them, when disordered, as nearly as possible, by changing the habits and circumstances of the patient, and afterward, or in connection, address remedies to the organs themselves. The stomach, liver, bowels, skin, kidneys, and uterus, should furnish their discharges in the most natural manner; and if they are not doing so, should be corrected by the most gentle means. If several of these organs are in a state of functional deviation from health, we should not expect to correct them all at one time, but alternate our attention between them; first, with our remedies influencing one, and then another. I insist here, with reference to the plan to be pursued, not to address all these organs, or even a large part of them, with medicinal agents at one time. There is no question, I think, but that complicated formulæ often nullify themselves by containing ingredients intended for the liver, kidneys, and skin, which ought all to act about the same time. We should act upon each of these alternately, in quick succession, if we think best; but let each organ feel the full impression of its remedy, before the blood and nervous energies are directed to another. In addition to this indirect way of increasing the tone of the nervous system, it is natural for us to look about for something that will act more directly. Our patient becomes so depressed, and suffers so much from terrible feelings of prostration, that her condition appeals to our sympathies for a more direct and immediate relief. If left to themselves, or the advice of injudicious friends, they almost always resort to stimulants, as whiskey, ether, chloroform, ammonia, &c. In some cases only are these temporary remedies advisable, and when used, they nearly always leave the patient in a worse condition than before they were taken. They are allowable only as necessary evils, and should be avoided when possible. These patients are usually depressed mentally also, and much good may be done by operating upon their minds. A physician who enters the room with a cheerful countenance, and a pleasant and gentle bearing toward the patient, and who engages her in conversation, first about her case, and afterward about some favorite theme, will do more toward temporarily relieving the great nervous and mental depression, than all the ether and ammonia the stomach can be made to bear. Earnest and kind assurances that her symptoms, though causing her a great deal of suffering, are not of a serious nature, and will soon subside, act generally as a good cordial to the spirit and nerves. In paroxysms of excessive nervous prostration, despondency, &c., I have seen the tonic influence of very cold air do a great deal toward relieving them. These paroxysms generally occur in close and often heated rooms—two conditions which should be removed. If it is cold weather, we should cover the patient to protect her, and let the frosty air—the colder the better—into the room, by opening *all* the windows and doors, and keep the room cleared of visitors. It will astonish anybody who has not observed the effect of a temperature near to zero, on those swooning hypochondriacs. A change almost immediately occurs for the better. If the air is not cold, it will still do much good to give it perfectly fresh to the patients in abundance. When able, they may be taken outdoors. This treatment introduces the natural stimulants, oxygen and cold, into the lungs, and brings them in contact with the nerves, and is more enlivening than medicine. How long the room should be kept open and cold, will depend upon the effect, but we should always, if possible, make these patients sleep in open, cold rooms. This is a very important item, which it will often require ingenuity as well as authority to enforce. These patients should live outdoor as nearly as possible, and be as much as they can on their feet.

Food.—Their food should have reference to the condition of the abdominal functions entirely, and be regulated by them. There is generally great intestinal torpor, which should be removed if possible.* Good cheerful company, travel—if the patient will not employ her body and mind in domestic pursuits —temperate and reasonable diversions, and, above all, time and patience, are requisite remedies. The affection is obstinate and chronic, and with the most judicious management will require time, if it does not vanish as the local treatment advances.

Nervous Excitability.—Connected with it often in some manner is great nervousness, excitability, irritability, or exaltation of all the nervous phenomena. This nervous irritability shows itself in great mental excitability, want of sleep, unreasonable agitation, restlessness, dissatisfaction; in short, in almost every phase of mental, muscular, or nervous excitement. There is also excitability of the different organs, with or without general nervousness, palpitation of the heart, nervous headache, local muscular contraction, &c. Successful management of these nervous and excitable patients requires a careful scrutiny into their general condition; the chylopoietic functions should be regulated in the most careful manner, the skin and kidneys should be attended to with great watchfulness. All that I have said as to general management in cases of nervous depression will apply to this kind of cases. As complete a revolution of the circumstances of the patient should be made as is practicable. From a life of ease, luxury, and absence of care, she should be, if possible, placed in circumstances requiring care, with muscular out-of-door exercise to the greatest extent she is capable of. If we cannot place our patients in situations which their cases require, we can send them on journeys that will demand exertion, calculation, care, and the deprivation of their usual domestic luxuries. The remark is frequently made that we must temper our remedies to the delicacy of the patients; and I am afraid that this injunction is misconstrued into the necessity of too great tenderness of treatment. The better rule is to make use of such means as will raise the patient from her state of delicacy to robustness. It is the delicacy of her constitution that causes her to suffer so much. This can be strengthened only by proper physical, moral, and mental training. The moral and mental condition of our patients when so very excitable should be at-

---

* See remarks on treatment of constipation.

tended to. Improper reading and society should be avoided, and social and literary habits should be reduced to great plainness and simplicity. Above all things, books and society should not interfere with regular rest, exercise, and outdoor exposure. As I have said before, this last should be as great in amount as can be borne, accompanied with active muscular exercise, as walking, and should be practiced in all weathers, sufficient protection being secured by enough clothing of the right sort. With regard to the use of medicine, it is a fact, that it is an exceedingly difficult thing to find any remedy that does not produce exaggerated and in most cases disagreeable and even injurious effects. So much excitability of the nervous system nearly always modifies the effects of remedies, and we can seldom predict the operation of any of them, nor can we determine the value of any until it has been tried. When tonics can be borne, they often very much relieve and sometimes entirely cure this great nervous excitability. Of the mineral tonics, probably bismuth, arsenic and zinc agree best. Iron is not frequently tolerated in any shape by these very nervous patients. Quinine, nux vomica, cherry and chamomile are the best tonics in these cases, but we must not be surprised if none of them are borne. Alcoholic stimulants in general agree with them, and are the best cordials for temporary nervous excitement, but should be conscientiously avoided when possible, as not a few, I am sorry to say, of most estimable and intelligent women have used them too much, and engendered an appetite that could not be denied. Opium, and in fact the narcotics generally, fail to have any good effect, but on the contrary disagree with the patient totally. This, however, is not always the case with opium, as it acts like a charm with some. In all it should be studiously avoided as deleterious in the long run, and there is danger of creating an appetite for it. We may the more readily be persuaded to omit the use of all these medicines, as their effects are temporary, while hygienic and regiminal remedies are permanent in their effects. The management of those cases of localized nervousness or unnatural excitability in particular organs, as palpitations of the heart, nervous headache, &c., is about the same as above, except that more attention to the stomach, from which they usually arise, may be necessary.

Some forms of nervous excitement are very much benefited by the bromide of potassium. Severe nervous headache, watchfulness, and neuralgic pains are often greatly relieved by this remedy. It should be given in full doses. For headache, from thirty to sixty grains every hour until relief is obtained. For watchfulness, the same quantity an hour before and at bedtime will sometimes procure a good night's rest. When given in full doses it should be dissolved in a large quantity of water, to prevent it from irritating the mucous membrane of the alimentary canal. I have sometimes succeeded in averting and preventing the return of the syncopal convulsions described under the head of general symptoms. One patient now under my care had been the subject of them for twelve months, having them several times a month. They had become so frequent and violent as to induce the fear of epilepsy, and had been treated with many remedies without material benefit. She has been taking the bromide of potassium for six months in doses of thirty grains three times a day, and during that time has had no convulsions. She is under treatment for endocervicitis. It remains to be seen, of course, whether this improvement be permanent, nor can I say how much of the amelioration may depend upon the treatment directed especially to the uterus. It is certain, however, that the "paroxysms," as she calls them, were improved immediately upon the commencement of the bromide treatment, and before I could reasonably expect benefit from the rest of the remedies.

We undoubtedly have a valuable means of relief from the pains attendant upon the condition of many of these patients in the hydrate of chloral, while it is often as prompt and positive in the relief it affords in sleeplessness and pain. So far as I am aware, it is not followed by the very disagreeable effects that result from the administration of opium and its preparations. It too should be dissolved in an abundance of water, to prevent it from producing local irritation upon the mucous membrane of the stomach, as it often otherwise causes vomiting or decided nausea.

Anæmia.—Anæmia, with its disagreeable concomitants, sometimes also calls for separate treatment. It would be an unnecessary waste of time and space to enter minutely into the general treatment necessary, where anæmia is the prominent and troublesome symptom. This condition calls for the same treatment found useful under other circumstances, and while it may not be entirely amenable to it, it will be very much benefited by the remedies indicated by the state of the blood. Iron, cod-liver oil, quinine, bitter infusions, and nutritious diet, with outdoor exercise to the extent the patient can bear, are the efficient remedies.

Plethora.—But we sometimes find general plethora instead of anæmia, a state in which there is actually an unusual amount and a too rich composition of the blood. I need not dwell upon this general state of the system, as the treatment is simple and familiar. The great fear is that, on account of the painfulness about the hips and legs, the patient may be too much inclined to an inactive life. On no account should this class of patients be allowed their ease; they must be urged to use up their surplus blood in active exercise, and the kind of exercise, next to the cares and labor of a household, best adapted to them, is walking. Every muscle in their body must be brought into action; every secretion must be kept free, and the mind ought to be taxed to continuous effort during the day by some useful occupation, while the strictest temperance, with reference to ingesta, should be their rule of living. Obesity and the troublesome and dangerous effects of plethora will be thus avoided, connected or unconnected with general plethora.

Local Congestions.—We sometimes meet with instances of violent, dangerous, and even fatal determinations of blood to particular organs, as the consequence of the general ill health which accompanies uterine disease, such as stupor, stertorous breathing, &c., indicating an oppressed condition of the brain, great dyspnœa, and sense of suffocation, showing congestion of the lungs. The treatment of these congestions does not differ from what would be appropriate under other circumstances of their occurrence, and consists in revellents, alteratives, &c. The most frequent, and perhaps obstinate, of the local congestions, are such as occur in the chylopoietic viscera, manifested by excessive secretion and discharges from the stomach and bowels. It is not uncommon for these patients to have suddenly recurring attacks of vomiting, cramps in the stomach and bowels, diarrhœa, and consequent great distress. Aside from the local treatment, we will be called upon to exert our skill against the exhausting and depressing influences of these attacks. It will almost always be found that such attacks are preceded by constipation, with scanty secretions, furred tongue, and other evidence of unhealthy secretions. By carefully correcting this condition, we may avert these painful and exhausting occurrences. The plan recommended and so much prescribed by Abernethy, will often palliate very much—viz.: six or eight grains of blue mass, at night, worked off by some saline cathartic in the morning of every fourth or fifth day. If there is more permanent diarrhœa, great care should be exercised in the choice of diet; the use of warm baths should be recommended, very warm clothing, and not much medicine, as the cure will depend upon the appropriate treatment of the local disease, instead of the treatment of the general symptoms. All these symptoms, except the diarrhœa, are apt to be moderate, and can be borne until the diseased uterus is cured; but there are two symptoms so very annoying, and which requires so much patience in the treatment, and exercise so much unfavorable influence upon the uterine disease, that I hope I will be pardoned by the reader for dwelling upon them more at length.

Constipation.—I allude to constipation and indigestion, particularly the former. I have already spoken of the deleterious influence of constipation, and I think I am justified in saying that, if disregarded, it retards the cure of chronic diseases of the unimpregnated uterus more than any other sympathetic affection. And I wish to warn the practitioner to be very particular in attending to this symptom. There is probably more tendency to costiveness in females than in males, chiefly owing to difference in habits. Sedentary life, confinement to close, badly ventilated rooms, are among the circumstances that bring on this condition. Irregularity of meals, late hours, deficient sleep, concentrated diet, imperfect mastication of food, all should be corrected, as any one of them alone will do harm, and all or any of these combined—and this is frequently the case—are very deleterious to the functions of the alimentary canal. But an inexcusable and very common custom of most females is making the act of defecation a disagreeable and procrastinated necessity, instead of a pleasant and punctual duty. The most trivial excuse—the presence of friends, a little cold, hot, or wet weather, being among strangers, or a slightly inconvenient distance from a proper place—will frequently be sufficient to limit defecation to once a week; then the act is performed in a hurried manner. It is amazing to know to what lengths this negligence is often carried. I have known two weeks to have transpired, frequently, according to the history of patients, without any attempt to relieve the bowels. Now this should be corrected by persistent method. The habit of eating from hunger at certain hours depends upon lifelong practice, and, when once established, cannot be changed without violence to many functions, causing urgent and repeated demands upon the system for a resumption of it. Regular bowels come from an equally long-continued habit of going to the close-stool at particular hours of the day. Years of negligence destroy the habitual regularity with which the bowels move; hence we should not be discouraged if the habit be not re-established without long perseverance. A new habit cannot be formed, nor an old one altered, without long and persevering effort in the right direction. We should, therefore, encourage a patient that is in earnest in her search after health, to persevere for months, years, and indeed her whole life if necessary, in going to her water-closet without fail, once every day, at a certain hour, as regularly as the clock points to it. This is indispensable to a correction of the bad habit of constipation. A very effective part of this regular endeavor, is to cause the mind to dwell upon the necessity for an evacuation, and the process itself, for at least half an hour before retiring to the proper place. It is not a difficult matter, with many persons, to create a desire in this way. Let no consideration of convenience enter into this punctual effort at stool. Arrived at the proper place, the position should be an easy one; no inconvenient strain upon any muscle should be allowed, and the patient should be possessed with an entire sense of leisure to perform the act completely. The value of all these considerations, where faithfully followed, is incalculable, and very few cases can long resist them. Without them, medicine will only temporarily relieve, instead of permanently curing, obstinate cases. I should caution against severe effort, or straining, as it is called; let time, patience, and gentle effort be the plan. Another matter of great importance, when an effort is made to have an evacuation, is to have the abdomen distended by ingesta. The patient should be instructed to eat plentifully of vegetable diet, such as by its bulk is calculated to produce fulness. If the patient go to the water-closet with a sense of fulness in the abdomen, success will be much more likely. Should the regular time for making an effort be soon after breakfast, which is undoubtedly the best time, and the meal has not been sufficient to produce a sense of moderate distension, a full glass of water will complete that condition. For the purpose of giving fulness and a sense of distension, various kinds of ripe fruit may be resorted to with advantage. In prescribing fruit for constipation, we should bear in mind that there are three indications fulfilled by it, some kinds fulfilling all, while others fulfil only a part of them. They are, first and best, distension; secondly, increase of secretion, on account of their acids; and, thirdly, increasing peristaltic action of the bowels by indigestible fibres, seeds, or rind. Ripe and mellow apples, without being divested of the rind, may be eaten in sufficient quantities to produce a sense of fulness, and this should always be at the conclusion of a meal, breakfast, for instance; the acids will increase the intestinal secretion, and the rind quicken the peristaltic motion of the bowels by acting directly upon the mucous membrane, and through it on the muscular structure. Very acid fruit, as lemon and orange, only produce their effect on account of the acids they contain. They are excellent as a part of the ingesta of patients whose stools are dry and hard and lumpy. Fruits containing an abundance of seeds, as figs, or of rind, as tamarind, &c., increase the peristaltic action without causing much secretion. By inquiring into the character of the stools, we will have a good guide as to the kind or mixture of fruits to be selected. There are kinds of diet, breads particularly, that act like these last fruits, and may be used in conjunction with or independent of them. Breads in which the bran, or hull of the grain, is contained in considerable quantities, are of this character. The Graham bread, as it is usually called, ordinary coarse brown corn bread, or wheat bread, are those mostly resorted to. When this kind of bread is used for constipation, it should be eaten at breakfast, dinner, and supper, in such quantities as the experience of the patient finds necessary. I have advised patients who could not use the coarse breads, to make what may be called bran crackers. A tablespoonful of flour, one pint of wheat bran, two tablespoonfuls of white sugar, and water enough to make them all into a pasty mixture, are the ingredients. This mixture is made into cakes, small or large, as may be wished, and baked in an oven until hard. When soaked in tea, coffee, or milk, they are not unpleasant. I have known patients benefited by swallowing certain seeds, with the rind whole. A tablespoonful of wheat grains, oats, barley, white mustard seed, &c., can all be used for this purpose, and are not more disagreeable than medicines. Another kind of diet which may be used to produce the kind of effect here aimed at, consists of the various small vegetables, as celery, radishes, pepper-grass, lettuce, asparagus, cabbage, &c. These may all be taken in quantities to cause distension.

In speaking of fruits, I ought to mention the berries as an excellent means, cheap and easily procured, to accomplish all the objects attained by other fruits.

Everything should be done by habitual effort, exercise, diet, drink, &c., before resorting to the use of medicines; because, as is well known to the patients generally, as well as to the practitioner, the more medicine taken the more will be necessary. They lose their influence, and the dose must be increased in order to produce a full effect. This is almost always the case. Notwithstanding this evil, we are often reduced to the necessity of using laxatives to overcome constipation. To a just and intelligent application of medicines in the treatment of constipation, it is indispensably necessary to make ourselves acquainted with the condition of the alimentary canal, with reference to its secretions and muscular powers. It will be found that there are sometimes great deficiency of secretion, and torpor or want of vitality of the muscular structure, or weakness of this tissue. The want of secretion may be in the upper portion, in which case the bilious color is wanting in the stools, or the small intestines may give out less watery material, and then the stools are less fluid, or even dry. The secretions may also be deficient in the lower portion, or colon; in which case the fæces will be scybalous, dry, and lumpy. The muscular torpor, from want of irritability, is more frequent in the colon or rectum than in the small intestines. When in the colon, there is increase in size of the lower abdomen, sense of fulness and hardness, and the fæces are expelled with great difficulty. If there is sufficient activity of the colon, but the rectum is torpid, large accumulations occur there, the pelvic distress is increased, and nervousness, general and local, is exceedingly annoying. Sometimes all these conditions are combined to render the case one of the most troublesome and difficult to manage. Mechanical obstruction by stricture of the rectum, formed by pressure of the uterus, may give rise to chronic constipation, which may become permanent and almost incurable; or the uterus, by lying on the bowel, and pressing it against the sacrum, often gives rise to costiveness, that can be removed only by correcting the position of that organ. It is not sufficient to know that the patient does not have regular operations from the bowels, but we must know why she is thus constipated. Whether on account of want of secretion, and, if so, of what secretion; whether it is attributable to general debility, combined with muscular weakness of the intestines, whether to lack of irritability of the intestinal tube and consequent torpor, and if so, where is this lack of irritability? Does it exist in the whole length of the canal, in the colon, or the rectum? Or whether there is obstruction from stricture in the rectum, piles, thickening of the mucous membrane, rigidity of the sphincter, or pressure from the uterus, bearing so heavily upon it. To give a laxative merely because it ordinarily produces a fecal discharge, is always unphilosophical, and sometimes exceedingly injurious in its effects. I think it is inattention to the exact state of the alimentary canal that makes constipation so often incurable. For constipation, attended with very dry, hard stools, showing a deficiency in all the secretions from the bowels, in addition to the course of diet, including acid fruits, &c., our object should be to administer such drugs as will most effectually stimulate to secretion. The various saline medicines are indicated. Sulphate of magnesia is a most excellent one; and a good way of administering it is in combination with sulphuric acid. From one to two drachms, or even half an ounce, given in combination with acid enough to taste somewhat sharply, will promote secretion along the whole of the small intestines, cause a large effusion of water, which will dissolve the fæces and render their evacuation easy and sure. In the morning, some time before eating, is the best time to take it. When there is reason to believe that the portal circulation is slow, and the liver furnishing less than its usual amount of secretion, some form of mercurial should be used with the salts. If the case is chronic and the constipation obstinate, we may give six to ten grains of blue mass in pills, at bedtime, every fourth or fifth night, and follow it with Epsom salts in the morning. A continuance of this alterative cathartic for four to six weeks, seldom fails to cause a change in the alimentary secretions. Sometimes it is better to give these cathartics nearer, and sometimes farther apart. We must judge of this more by the susceptibility to the constitutional influence of mercury than anything else. It is almost always the case that this very scanty state of the secretions is accompanied with an impoverished state of the blood; hence iron in some shape will be beneficial in most cases. If there is much debility, a long course of tonics will be indispensable. It may often happen that this scanty condition of the secretions is attended with debility of the muscular fibre of the intestinal canal. When this is the case, we must add to the above treatment that which is applicable to this kind of intestinal torpor, which I shall now consider. Before doing so, however, I will remark that several other salts will answer as well, and sometimes even better, than sul. magnesia. The kind of tonics which are most effectual in debility of the muscular structure of the intestinal canal are such as give general strength, and it is mostly desirable to combine them with special tonics. The latter are rhubarb and nux vomica. These have always seemed to me to have a special tonic influence upon the intestinal tube, and, when properly given, to increase the susceptibility to their own action. The rhubarb, although an alimentary tonic, induces less susceptibility to its own influence than the nux vomica. The best way to give the rhubarb is either in the root, without pulverization, or in the extract. When given alone in the root, the patient can take a little, twice a day, by chewing it, and after mixing with the saliva swallowing it. A little experience will enable the patient to judge of the right quantity, which she can repeat as often as it is required. When the rhubarb is taken this way, she may also take a solution of sul. ferri and strychnia, in water, one grain of the former to one-sixteenth of a grain of the latter.

A formula that is very simple and effective is as follows:

 ℞. Strychnia, ... gr. j.
  Ferri Sul., ... gr. viij.
  Acid Sul. q. s.
  Aqua, ... ℥ij.

Mix. Make solution. One teaspoonful three times a day after eating. Sixteen grains sul. quin. may be added to the above formula to advantage. Or,

 ℞. Strychnia, ... gr. j.
  Ext. Rhei, ... ℈iss.
  Sul. Ferri, ... gr. x.

Mix. Make sixteen pills. One to be taken once, twice, or three times a day, as may be found necessary.

I have often succeeded in overcoming this constipation or debility by giving one grain of sul. quin. with five grains of powdered nux vomica after each meal. Or the same amount of nux vomica, with iron by hydrogen, two grains, after eating each time. It is usual to use aloes in the constipation of uterine diseases; but I have found very few cases with which this drug did not disagree. But there is a torpor of the intestines where general tonics cannot be borne; where, in fact, there does not seem to be any general debility, there is only a want of susceptibility to the stimuli which ordinarily arouse them to action. The secretions color the fæces properly, and give them sufficient moisture; there seems to be no fault in their appearance, consistence, odor, or other character whatever. They are deficient only. The patient may be plethoric and florid, her general muscular strength sufficient, and her blood, so far as we can judge, good in composition. Special tonics and stimuli are indicated in such instances, and they alone should be used. Such measures should be adopted as will arouse the muscular action of the intestines. Nux vomica, in five-grain doses, with the rhubarb extract or without it, or the strychnia in solution, in doses of sixteenth to twentieth of a grain, constitute our most valuable medicinal appliances. This is the kind of constipation that is most benefited, and is most amenable to a persevering regiminal and dietetic course of management, such as I have above endeavored to give. In addition to the rhubarb and nux vomica treatment, we may get some good from external appliances and manipulations to the walls of the abdomen. The most valuable, when gently, perseveringly, and methodically applied, is what is understood by the term kneading. The colon is the torpid portion in most cases of this sort of constipation. The process of kneading consists in handling it so as to stimulate its fibres directly. One plan is to grasp it with the hand, and squeeze it from one end to the other. We should begin at the right groin, and with a knowledge of the position and direction of it, grasp it with both hands at this point, then a little higher up on the same side, and then a little higher, until we reach the right hypochondriac region. We should then follow it across the abdomen to the left hypochondriac region, and thence down to the left iliac. Or, we may double our hands as bakers do when kneading their dough, and standing over the patient, press with the knuckles of both hands, first in the right iliac region, and imitating the process of kneading, pass slowly from this to the right hypochondriac, thence across the abdomen and down, as before directed. If we trust this process to a non-professional attendant, we should be sure to show him how to do it, as it is important that it should be done right. When this process of kneading or squeezing the colon is first instituted, it should be practiced with the utmost gentleness, but the force and rapidity of motion may be increased until great freedom may be made use of. It should be resorted to a short time before retiring to the water-closet, say half an hour. Some patients find an efficient laxative in what they sometimes call a water-compress, applied to the abdomen over night. It is made by doubling a napkin several times, so as to make a thick compress, large enough to cover the entire abdomen anteriorly. This is saturated with water, and, after being placed upon the abdomen, covered with a roller or bandage so as to keep it in place. It is thus allowed to remain from the time of going to bed until the time to rise in the morning. I think this water-compress is best adapted to cases in which there is a deficiency of secretion in the intestinal tube. A bandage, or, what is better, a roller applied lightly enough to press the walls strongly upon the contents of the abdomen, frequently stimulates them to proper action, both as it respects secretion and peristaltic motion. When it is determined to use the roller or bandage for its stimulating influence, it ought to be applied upon rising in the morning, or, what is perhaps better, immediately after breakfast. This bandage should not be worn constantly, nor even many hours in the day. From the time of rising until two hours after breakfast, or from breakfast for three hours thereafter, will be long enough. The constant use of the bandage would but increase the evil—lax abdominal muscles—for which it is advised. Before leaving this part of the subject, I desire to say, with reference to the freedom with which I have advised nux vomica to overcome intestinal torpor, that in all cases we should remember its effects are cumulative, and quite a difference of susceptibility to its influence is manifested by different persons, in consequence of which the patient should be watched, and the dose graduated to the least quantity necessary in the case. Although I have given nux vomica and strychnia for a considerable length of time to a great variety of persons, and for several weeks together, I have never seen anything more than slight inconvenience from it in the shape of nervous startings. Very rarely we meet persons who cannot take it at all; it disagrees with them as soon as they commence its use. There is another species of intestinal torpor of a very obstinate character and very distressing to the patient; I mean a lax, torpid rectum; so torpid as to allow the fæces to accumulate in large quantities, and cause great inconvenience from pressure. To such an extent does this collection sometimes go as to press the posterior walls of the vagina forward and protrude it between the labia. The first indication in such cases is to dissolve the fecal mass and discharge it. Various kinds of injections are useful for this purpose, warm oil, warm water, &c.; but one which I have seen do much good is composed of one ounce of fresh ox-gall and four ounces of warm water. This composition dissolves the fæces very readily, and the fresh bile stimulates the intestine to their expulsion. The evacuation, of course, will give only temporary relief, and there remains the most important indication, that of giving tone to the bowels, with a view of preventing the accumulation in future. This is difficult, and in some instances of long standing quite impossible. Much good can be done in nearly all cases, however, and we do not discharge our duty if we do not try to relieve when we cannot cure every case. Cold water thrown into the rectum once or twice a day, in small quantities—eight ounces—is always good, without some special reason to the contrary. There are generally two indications to be fulfilled in these cases—relaxation of the sphincters and restoring the tonicity of the proper rectal fibres. It is a singular fact, which I think I have observed, that the sphincter muscles increase in strength with the advance of age; this is one of the causes why the fæces are voided with more difficulty in old persons. To give tone to the rectal muscles, astringent injections have been recommended and extensively used; but in my practice they have been almost uniformly useless, many times injurious, and always disagreeable. They dry up the secretions, an evil not to be compensated for by any other effect; they do not, so far as I can judge, cause contraction of the muscular fibres, but they are very apt, if persisted in for a length of time, to cause inflammation. I have derived more benefit from tonic suppositories and injections than from any other kind of medicinal treatment. A suppository of twenty grains extract gentian, or five grains sul. quin., ten grains extract cornus Florida, or a mucilaginous suspension of any of these introduced into the rectum every night at bedtime, and retained, if possible, until morning, are good tonics and eligible modes of using them. It will be necessary, to secure the retention and efficient contact of these tonics, to first empty the bowels with ox-gall and warm water, and afterward introduce them with as little irritation as possible. The quantity of mucilaginous material should not exceed two ounces. The tonic treatment of this kind must be varied, taking first one tonic and then another, in first one form and then a different one, and must be kept up for a long time to do much good. We cannot be too careful, in all our treatment, to avoid anything to which the rectum shows any sensitiveness. When it becomes tender and sensitive, we should at once desist until all of this has subsided before we are justified in beginning again. It too frequently happens that both the physician and patient become discouraged, and desist before the remedies have had a fair trial. Is there anything that will relax the sphincter ani? I am not aware that any means operate with efficiency in this direction; but I have used, in a few instances, with apparent benefit, the ointment of belladonna, made by mixing the extract with lard. I apply it to the anus externally upon going to bed at night, and think that it promises decided encouragement to continue, until the question against or in favor of its usefulness is fully determined.

R. Ext. Belladonna, ... ʒij.
Ung. Simplex, ... ʒj.
Mix. Make ointment. This is a good formula. The parts to be well smeared with it at bedtime.

This application certainly removes the irritability of the sphincter, which causes it sometimes to resist the extension of the fæces. As I have before remarked, there are cases in which this relaxation cannot be cured; we are then compelled to resort to palliatives, and we must be careful to palliate intelligently. We are to give the weak rectum artificial support, to enable it to retain as near as may be its ordinary size. This can be done only through the vagina. An air or sponge pessary introduced into the vagina, so as to press the rectum against the sacrum, and thus diminish its capacity, will prevent the great accumulations from taking place, and in that way prevent one source of great inconvenience. Dr. Hodge recommends the globe pessary for this condition of the rectum, which answers very well in many cases, perhaps in the majority; but each case must be studied with reference to its own peculiarities, and the shape, size, and consistency of the pessary adapted to it. When our object is palliation alone, there is no objection to wearing the pessary all the time, but if it is used to palliate what we believe to be a curable case, we ought to use it intermittingly, and the patient should not wear it at night especially. It would probably be better in a majority of the cases to introduce it before rising in the morning, and allow it to remain until noon. One thing I think essential in the size and position of the pessary, and that is, that it does not compress the rectum below its natural capacity; there should be room enough for an ordinary amount of fæces in it, lest it become a source of obstruction, which it will do when larger or improperly placed. As will be noticed, I have omitted to say anything of enemata in constipation, from inactivity of the colon or upper portion of the alimentary canal. As an occasional means, injections operate well; but, like other laxatives, when used for a length of time they lose their influence entirely. If we determine to use injections as an habitual laxative, by proper changes in kind and quantity, we may prolong their efficacy very much. To a person unused to them, half a pint of cold water will act very well. When the bowels fail to respond to this quantity, there ought to be an increase of two or three ounces, and then that amount used until its effects are not satisfactory, when a few ounces more should be added, and so on we may increase the amount until the quantity becomes intolerable. When this is the case, we may order half a pint of water with a drachm or two of common salt, chlorate potassa, or nitrate of soda or potassa. We should increase the quantity of water or strength of solution, or both, as the susceptibility of the rectum decreased, until we cannot carry either farther. After we have thus obtained as much good from injections as we can, it is sometimes expedient to use suppositories as laxatives. Suppositories are made of laxative medicines or of any other material. Compound extract colocynth or other purgative extract may be used; or we may inclose in some of the extracts a dose of the podophyllum, or any of the purgative resinoids or alkaloids. These should be retained until absorption takes place. The common suppositories of soap, tallow, and wax, sperm, stearin, &c., are of the second kind. It not unfrequently happens that the above modes of using injections and suppositories may be alternated very profitably, the full effects of each being experienced upon their resumption after having used the other for a time. But some persons cannot use injections; the rectum is too sensitive, and attempts to do so induce so much irritation that they must abandon them. In such cases suppositories are out of the question.

I have elsewhere shown that the uterus, by its wrong position, sometimes presses upon the rectum and obstructs the passage of the fæces. This may be effected by retroversion or prolapse. The indication, of course, is to restore the uterus to its proper place, and as I shall have occasion to speak elsewhere of these difficulties (malpositions), I do not think it necessary to more than mention them here.

Indigestion is another very troublesome condition among the many which attend uterine disease, and it will demand much of our attention. It would not be profitable to dwell at any great length upon this symptom, as it will become the duty of the physician to study each case separately. Attention to the bowels, keeping them perfectly regular, will very much alleviate if it does not cure most cases; but sometimes we find the stomach very seriously disordered when the bowels are perfectly regular. In such cases we should inquire into the alkalinity and acidity of the urine as a good index for the administration of medicine. If the urine is highly alkaline, acids and bitters are indicated and will be well borne; if the urine is highly acid in its manifestation, alkalies must be used—liquor potassa, lime-water, soda, &c. In the former case, animal diet may be tried; in the latter, vegetable diet, as likely to be good palliatives, and under proper circumstances curative. The indigestion, like most other symptoms, however, will be obstinate and generally incurable until after the local disease is cured. It may be inferred from what I have already said, that I consider the general treatment, as I have endeavored to sketch it, of secondary importance, and the local as the essential treatment; but wishing to be perfectly clear on this point, I will reiterate what I have already said in regard to the objects of general treatment. They are—1st, To palliate the general condition of the patient before and during local treatment; 2dly, To aid local treatment in effecting the cure; and 3dly, To cure the effects which may remain after the local disease has been removed. I do not believe that a radical cure is ever effected by general treatment alone.

CHAPTER XI

LOCAL TREATMENT.

Baths.—The local treatment of inflammation of the cervix uteri is made up of several therapeutic items, varying according to the intensity, quality, and seat of disease. Of these there are, however, a few that are applicable to almost all cases; hence their description, modes of use, &c., may be considered before going farther. Baths, injections, and some minor remedies are of this kind. Water, when applied to the surface, is purely sedative in its effects if it is of the temperature of the part on which it is used. If the bath is partial, the sedative influence is for the most part confined or limited to the part to which the application is made. So with injections per rectum or vaginam. They soothe the parts contained in the pelvis. If the water is warmer than the part of the surface bathed, the effect is stimulant; if it is colder, by virtue of the physiological action brought into play, it is first sedative and then stimulant. The circulation and nervous influence of the vagina, for instance, when the cold water is first thrown into it, are depressed, but very soon after its evacuation, or withdrawal, the vessels become excited to increased circulation of blood, and increased heat takes place and the nerves become more sensitive. In all these respects baths and injections act alike. The injections are internal baths; the uterus is bathed through the vagina by injections. But the effects of baths and injections may be modified by containing medicinal substances. They may be rendered more stimulant or more sedative, or be even made to possess other qualities by impregnation with medicines: one of which in very common use is astringent in character. Another mode of using water and applying it, either simple or impregnated with medicine, is, to wet a cloth or a sponge with it and bind it to the surface, or introduce it into the vagina. Several thicknesses of cotton cloth applied to the abdomen and impregnated with water is what is called the water compress; and often when allowed to remain in contact with the skin for several hours it produces considerable excitement, and if persisted in for days, will cause first a vesicular, next a pustular, and finally a phlegmonous eruption. The way to render them effective is, after applying the wet cloth compress to cover it over with oil-silk, and then confine the whole with a bandage or roller, with a view to prevent evaporation. Sponge introduced into the vagina impregnated with water holding medicine in solution is a common way of affecting the uterus. I do not design giving an extended view of baths or their application and modus operandi, but so much aid is occasionally obtained by the use of them, that I cannot refrain from speaking of the application of some forms of them to diseases of the uterus. The baths most applicable in inflammation of the cervix uteri and most commonly used is the sitz or hip-bath. They are intended to allay the inflammatory irritation and pain. It is often the case that there is a great deal of suffering from pain without much inflammatory action in the parts; in these cases a sitz bath will often give great relief. In many instances the efficacy of the bath may be enhanced by having the patient introduce a speculum while in the water, so that it may pass up the vagina to the neck of the uterus and thus directly affect the part diseased. In cases of medicated sitz baths the organ may thus receive the full benefit of the saline, anodyne, or other medicinal impregnation. The common glass tube will do very well for this use, where we wish only to bathe the neck of the uterus; but if we wish the fluid to come in contact with the vaginal walls and remain there for a considerable time, the wire speculum is the best. While speaking of the use of the speculum in this way, I may mention that a very efficacious mode of applying medicated washes without the bath to the cervix uteri or vaginal walls, is to have the patient lie upon her back, introduce the speculum, and then pour the fluid into it. By remaining in that position she can retain the contact of the medicated solution as long as desirable. Ice water, ice, astringent powders, or almost any form of substance may be applied and retained in contact with the os and cervix uteri with great advantage in this way. This mode of using remedies is particularly useful in bleeding fungus or vascular tumor of any sort.

Hip Bath.—The sitz bath, when a patient is suffering with the pain and heat of uterine disease, may be used as often as necessary, twice a day at least; but three, four, or even a greater number of times will not be too often, when they are found to be soothing and useful. We may extemporize a hip or sitz bath, by putting water in a common washing tub; but the cheap tin vessels made for the purpose are within the command of almost all persons. There should be so much water that when the patient sits down in it, the whole pelvis will be covered.

Temperature of the Bath.—What should be the temperature of the bath? The patient's sense of comfort, or discomfort, from its use, should be our guide in this respect. We should seek a temperature that is comfortable and soothing to the patient while in the water, and that leaves no sense of discomfort. The baths are intended for, and should add to, the comfort of the patient; when they do not do this, they should at once be discontinued. As a general rule, I advise my patient to take tepid water for her first bath, and then gradually use them cooler until they are cold, unless they become disagreeable in some respect; if they do so, to continue them tepid. The colder a bath is, the more good it does, provided it be comfortable. The time for taking it may be regulated by the convenience of the patient, and the necessity for it, with the view of allaying pain, heat, &c.; probably in the majority of instances, the most advisable times for taking it are upon rising and retiring. The length of time the patient remains in the bath should also be regulated somewhat by their effects. If the patient remain too long in the water, it will debilitate her, particularly if there is considerable water and the bath is frequently repeated; on the other hand, if she does not remain long enough, she will not derive any benefit from it. She may try remaining in it fifteen minutes, if she does not find herself very much relieved before that time, and she ought to be governed in her use of subsequent baths in this particular by the effects of the first few trials. While in the bath, the intended temperature of the water may be kept up by adding hot water from time to time. The hip bath is used almost wholly with reference to the local disease, but when general baths are required, it is usually for the relief of some attendant general symptom.

Shower Bath.—The shower bath may be used as a roborant exciter of the circulation, if, upon trial, it can be borne, and produce a good effect. Some patients think they are very much benefited by the shower bath, and say they cannot do without it.

Sponge Bath.—The sponge bath is useful in causing a tonic and soothing reaction upon the surface. Neither of these can be tolerated by very feeble patients. The cold or tepid sponge bath, administered at bedtime, not unfrequently soothes nervous irritability, and enables restless persons to sleep soundly. I have not used baths in any other form than these, but when used as I have here indicated, I have seen such pleasant results from them, that I cannot refrain from recommending them.

Injections.—Injections are applicable to almost all cases of inflammation of the cervix uteri, do a great deal of good, and are believed to be sufficient to cure some cases. As I have before said, they may be used as internal baths, to get the influence of water and temperature on the vagina and uterus, for the application of medicinal substances to the mucous surface of this cavity and viscus, and also as detergents, to wash the vagina of all substances that should be removed from it for purposes of cleanliness. In some one of these forms injections may be used in nearly every sort of cervical inflammation. The simple injection of water may, and ought to be used by all females who have inflammation of the uterine neck. The medicated injections can be useful only in cases where the inflammation is within reach of them, as when inflammation affects the mucous membrane of the vagina, or the membrane covering the external surface of the vaginal portion of the cervix. For obvious reasons, injections containing medicines can hardly do any good, by virtue of the solution, when the inflammation is situated inside the cervical cavity. Vaginal injections cannot reach the seat of disease. I have not used intra-uterine injections, as I think there are less hazardous modes of conveying medicines into the cavity of that organ. I should not discharge what I consider a duty in this respect, if I did not condemn the use of the intra-uterine injection. This method of reaching disease in the body of the uterus, has lately been so strongly recommended by a number of eminent men in the profession, that it will undoubtedly be more extensively resorted to than it ever has been before. I think this is unfortunate, and believe that sufficient facts have already been accumulated, showing the suffering and danger resulting from it, to condemn it, without subjecting this class of patients to the ordeal of a new trial. I believe a great amount of harm has been done, and that much more will be perpetrated by it. My own observation was conducted under a conviction of the correctness of its philosophy, and with an earnest desire to avail myself of every good means of curing my patients. The result of such trials as I have made is, that they have none of them been useful when used for any other purpose than checking hemorrhage in cases of abortion or uterine fungus. I think, also, that they are unnecessary, as safer and more efficacious methods have been devised for treatment of the mucous membrane of the corpus uteri.

Manner of using Injections—Kind of Syringe.—The efficacy of injections depends very much upon the manner in which they are administered, and the kind of instrument used. The essential quality of a syringe is its capability of receiving at one end and discharging at the other perpetually, so that any quantity of water may be used without withdrawing and reintroducing the pipe. A large number of forms of syringe have been invented, but for convenience, that form is, I think, preferable which has a vulcanized rubber hollow ball mounted in the middle of a long flexible tube; by pressing on this ball, and relaxing it, the water is drawn in at one end and forced out at the other. A pewter, britannia, or ivory tube delivers the water into the vagina, and by its length may be made to convey it to the uppermost part

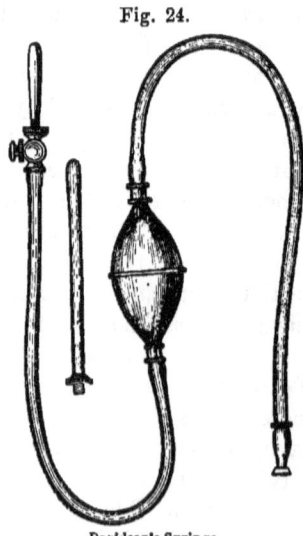

Fig. 24.

Davidson's Syringe.

of that cavity, and thus completely wash the whole of its walls. A siphon may be made to answer the same purpose, by having the fountain high enough to give some force to the current. Should the patient use a syringe of the above description, she may sit over one vessel, and have the water in another in front of her. By inserting one end in the vagina, and the other in the vessel of water, the whole of it may be made to pass through the vagina and fall into the vessel beneath her, and thus do away with the inconvenience of undressing.

An instrument is now made and on sale by Messrs. Burbank & Co., Boston, that in many instances is an admirable substitute

Fig. 25.

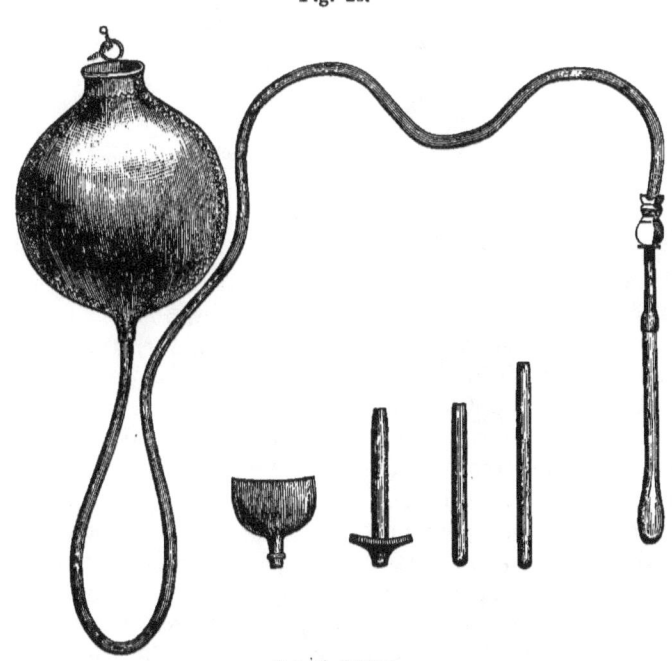

Fountain Syringe.

for syringing; they call it the fountain syringe. It is an India-rubber sack with a long tube depending from it. The sack is filled with water and hung up several feet higher than the patient, the tube is then inserted in the vagina, and, by means of the thumb and finger, the flow is regulated to suit the circumstances. I give a wood-cut of one of them that will convey a correct idea of them.

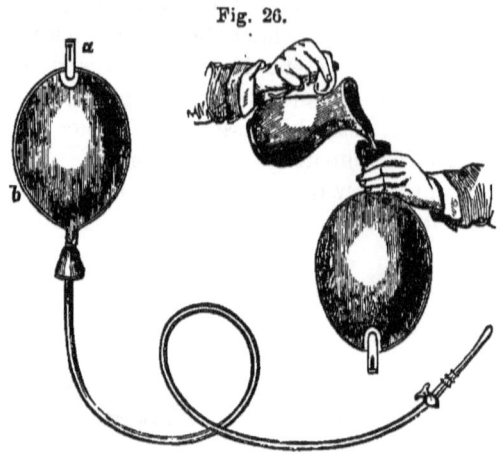

Fig. 26.

Quantity of Injection.—The quantity necessary to be used in an injection will vary very much in different sorts of cases; if water alone is to be used, and we wish to get the sedative influence, the quantity must generally be large—that is, from one to eight quarts; if we wish to stimulate the uterus with very warm water, a large quantity will also be necessary; if we wish the injections cold, it is better not to use so much.

Medicated Injections.—The medicated injections also should be large or small, according to the effect we wish to produce, and the strength of the solution. A pint, or at most a quart, will be sufficient for astringent injections. We often use anodyne injections on account of their soothing influence upon the sensitive parts. As a general rule, anodyne injections need not to be very large, say a pint, or less, but the patient can continue passing it through the vagina until its effect is attained. This may be done by using only one vessel, pumping from and allowing it to fall into the same. Frequency must be determined also by the object of the injection. Simple water injections can be used more frequently than medicated ones, and anodyne more frequently than astringent. The simple injections, if they afford relief, may be used from three to six times a day, or oftener; narcotics three or four times, or oftener, owing to the urgency of the symptoms requiring them, and the good they are found to do.

Astringent Injections.—Astringent injections ought not to be made use of, as a general thing, oftener than twice a day, and in some cases to which they are applicable, this is entirely too often. Of all the vaginal injections used, the astringents are most commonly resorted to, and are productive of most good.

Modus Operandi.—When an astringent is thrown into the vagina, the first effect is to coagulate the mucus, pus, or blood, contained in it; after this, its contact with the mucous membrane becomes more intimate, and its influence is exerted upon the capillary bloodvessels, and the glandulæ or crypts. The vessels are constricted in size, and circulate less blood, and the calibre and functional activity of the crypts are diminished, and slight congestions and inflammations are for the most part cured, or at any rate benefited. When the vessels are circulating too much blood, and the muciparous apparatus furnishing too much secretion, this astringency is desirable. We ought not, with certain exceptions, to use astringent injections when there is no hypersecretion from the mucous membrane of the vagina or cervix uteri, nor an ulcerated or inflamed surface with which the solution can come in contact.

Frequency of Using.—The frequency which they may be used must be indicated by observing these two effects, and the dryness more particularly. I think we may lay down a rule for repeating them, like this: never repeat an astringent injection while the vagina is dry from the effects of a preceding one. We should, after obtaining the full astringency of an injection, in the stoppage of a leucorrhœal discharge, wait until the mucus again renders the mucous membrane moist. It will be found, very often, that this requires twenty-four and even thirty-six hours to take place. A disregard of this direction will sometimes induce an increase of inflammation, and give our patient great inconvenience. In fact, too long a continuance of astringent injections is apt to cause vaginal inflammation.

Alternate Different Astringent Remedies.—I think, however much an astringent may be indicated, that the same article ought not to be used more than twelve or fourteen consecutive days, and should then be alternated with another one of the same class, or simpler ones. This last I generally prefer. A permanent dryness of the vagina after any one astringent, should preclude the use of that article at least, and cause us to try another, and so on until we get one that will agree with the case; or else we must abandon all astringents, and fall back upon simple water. To get the full benefit of a medicated injection it should be preceded by one of simple water, in order to wash out the superabundant secretion in the vagina.

Temperature of Injections.—I know of no better rule to govern the temperature of injections than the comfort of the patient. After a trial of tepid, warm, cool, and cold, let the patient suit herself by the effect they have upon her. Any temperature that is disagreeable should be avoided. The extract of opium makes a good anodyne injection. Five grains to a pint of tepid water, used for ten minutes, a quarter or half an hour, will often allay pain, arising from inflammation within the vagina, very readily; or one grain of extract of belladonna may be used in the same way. In fact, we may choose among the narcotic extracts, remembering that the solution must be impregnated with at least three doses of the medicine. Among the astringents, alum is the most common, the most useful and efficient astringent. It possesses the advantage of having no poisonous ingredient in it. As Dr. Bennett has taught us, it sometimes produces severe inflammation; but this is doubtless owing to the inconsiderate use of it, and arises from its very efficacy in suppressing the vaginal secretion. One drachm to the quart of water, tepid, cold, or warm, as the patient may desire, is perhaps the strength of solution that will most commonly agree well; but in this the patient should be governed by the sensation it leaves behind. There should, at first, be a sense of dryness, quite obvious to the patient, which should pass entirely off in less than six hours; much better if it is entirely gone in two hours after the injection is administered. If this sense of dryness is perceptible, we should not allow the patient to use an injection for several hours after it is gone; and the longer it continues, the longer should be the interval between them. If it last six hours, the interval should be twenty-four; if two hours, the interval should be twelve; if it last twelve hours, it should be discontinued, as it will most likely do harm. Another good astringent is sugar of lead; this is, perhaps, next in efficiency to the alum. Double the quantity may be dissolved in the same amount of water. I do not like sulphate of zinc, although highly recommended. Thirty grains of it may be dissolved in a pint of water, as an astringent injection. The sugar of lead, or zinc, ought not to be continued as long as the alum. Some of the vegetable astringents are often used to good advantage; strong decoctions of oak bark, rhatany, kino, or solution of pure tannic acid. This last is an admirable astringent, not less efficient than the metallic, but also less injurious. It can be used of the same strength as alum, or even in double that strength, if desired. Injections and baths ought to be suspended during the time for menstruating; if tepid and simple, they probably do no harm at this time; but if cold or astringent, they are pretty sure to interrupt, more or less completely, this flow. Almost every practitioner that has had much experience in the treatment of uterine diseases has a favorite injection. I am disposed to adhere to the simpler forms, seeking rather for correct principles by which to be governed in administering them, than for great variety of substances.

Accident in Injection.—There is one annoying, and sometimes to the patient alarming, little accident that occasionally occurs during the reception of an injection in the vagina. Suddenly, while injecting the fluid, she is seized with severe cramping pain in the hypogastric region, which radiates to the back and hips, down the thighs, and sometimes over the whole abdomen. She becomes sick at her stomach, is attacked with rigors, and her feet and hands often become cold. This pain continues, with exacerbations and remissions, for several minutes or hours, and when it subsides, leaves a sense of soreness, more or less considerable, corresponding with the severity of the attack. As the chilliness and rigors of the first few moments subside, there is reaction; the patient becomes warm, and sometimes decidedly feverish. In all cases in which I have witnessed these symptoms the patients were using a syringe, in the end of which, within the vagina, were several perforations, some on the side of the bulb at the end, and one at the very extremity. I think that one of the perforations had been accidentally placed in apposition with the external os uteri, and as the water was forced through this perforation, it entered the cavity of the cervix, and passed through it into the cavity of the body of the uterus, inducing the first shock, and the pains following it were caused by the spasmodic attempts on the part of the uterus to expel it. Although I have, in a large number of instances, been called upon to witness and prescribe for these symptoms, I have not seen them proceed to dangerous extremities. I think these are cases of injection into the womb; and, in this respect, they constitute my whole observation. An opiate injection per rectum, fomentations over the pubis, and quiet, are all the remedies I have found necessary. And often the symptoms subside so soon that I have not been under the necessity of prescribing at all.

We occasionally meet with patients who cannot use baths or injections. In these cases it will be found, almost invariably, that this inability arises from their producing an exaggerated effect. If it is simple tepid water used for the bath or injection, its results are too sedative. The bath debilitates the patient, instead of simply soothing her. I have seen a single tepid bath prostrate a patient so that she would have to lie in bed for several hours before its effects wore off. A cold bath induces chilliness and permanent coldness, and reaction is not established; the system recovers from its effects only after a number of hours, and that slowly. Hip, sitz, or general baths may produce these effects, and when they do so, should be abandoned as injurious. Other nervous symptoms, as difficulty of breathing, nausea, dysuria, &c., also occasionally seem to be the effects of baths. It is singular that some patients are so susceptible to the depressing effects of water that injections debilitate them very rapidly, and they are obliged to abandon them on this account. Cold water, as an injection, not unfrequently causes general coldness. But it is the medicated injections that most frequently produce an exaggerated effect. Alum injections, even when the solution is weak, with some patients produce such disagreeable and constant dryness, and sense of heat, as to make them quite intolerable. And the sensitiveness of the vagina becomes so great that some patients are forced to cease the injections of alum wholly. The same objections apply to other astringents to a less degree, and the consequence is, that however baths and injections may seem to be indicated, in the cases where idiosyncrasy renders them so objectionable, we must forego their use entirely.

Should they be used in Pregnancy?—Is pregnancy an objection to the use of local baths and injections? I think not, with proper care. A hot bath about the hips would be objectionable; a very cold bath that might cause much of a shock, or internal congestions, would not be advisable; but plenty of tepid water, and even cool water temperately used as baths, give the pregnant woman great comfort, and cannot generally be followed by any bad effect. Injections may be used with less caution than baths. The caution which we would administer to all is, that they should not be copious. In pregnancy the patient ought not to use more than a quart at one time. The injections should always be tepid or cool; not very cold nor very warm, lest they stimulate the muscular, vascular, or nervous system of the uterus too much, and induce hemorrhage, or provoke contractions. Both of these effects, I think, I have known produced by such injections; the cold causing contraction and expulsion; and the very warm, hemorrhage and death of the ovum. Strong astringents should also be avoided. Much comfort may be derived from anodyne injections, when there is neuralgic suffering about the uterus or vagina, during pregnancy. Cases of superficial inflammation, and even early ulceration of the vaginal portion of the cervix, may always be benefited by injections, baths, and the general treatment which I have heretofore detailed. In fact, most cases, if not all, where there is no idiosyncratic objection to the baths and injections, will be very much benefited by them. When, however, the disease has been of long standing, or extends between the labia of the os uteri, or into the cavity of the cervix, these will only slightly benefit it. We must then seek for something that will more profoundly influence the nutritional changes, and the vascular and nervous tissues of the parts.

The introduction of anodyne, astringent, and alterative ointments, pessaries, and powders, may be resorted to with much profit in many instances. The small instrument called the suppository syringe will enable the patient to place ointment in contact with the uterus very conveniently. Ointments made with opium, belladonna, hyoscyamus, cicuta, tannic acid, or other astringents, mercury, iodine, in fact almost any substance used to exert an influence locally, may be made into

Fig. 27.

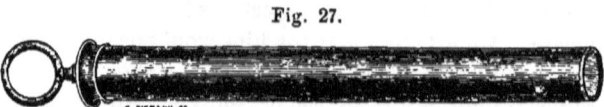

ointment and thus introduced. The powders of many of these articles may be deposited in the vagina in the same way. And the medicated pessaries made by mixing the medicine intended to be used with cocoa-butter, and passed up to the os uteri through a glass speculum, either by the patient, her attendants, or the physician. In using the narcotics in the vagina in the form of ointment or pessary, we can safely use double the quantity given by the stomach. The ointment is absorbed slowly, and consequently it requires some time to effect much by it. But the powders act much more readily. Morphia thus introduced will sometimes act with great promptitude, and the powder of tannic acid is a very efficient astringent used in this way. The absorbing power of the vaginal mucous membrane is decidedly less than that of the rectum. It takes a longer time and more of the medicine to affect the system through this cavity. Possibly this may be to some extent on account of the more ready escape of substances from the vagina; but I think also the membrane does not take up substances so quickly. From this fact, injections or suppositories per rectum will often do more good in allaying pain, especially than when used per vaginam. A few drops of strong solution of sul. morphia in the rectum acts very promptly. Dr. Greenhalgh and others use cotton pessaries medicated, per vaginam. The cotton is prepared by immersing in a strong solution of the medicinal agent to be employed, and thus impregnating it, afterward drying, and then using it. Still another method of making local applications to the upper part of the vagina is to envelop the medicines in a sack of thin cotton or linen goods, and pass it up to the cervix, and let it remain there until the astringent, or whatever may be contained in it, is dissolved out, and exerts its influence upon the parts. The patient can use this kind of application without assistance.

Principles that should govern us in choosing the Kind of Local Treatment.—The local remedies I have been in the habit of using more frequently, and, in fact, almost exclusively, are the various depletory measures, nitrate of silver, tannin, acid nitrate of mercury, nitric acid, and caustic potassa. Of these, the nitrate of silver is most frequently used. In fact, it has so generally answered the purpose in my hands, that I look upon the others as substitutes, and to be used only when it disagrees or fails. This, of course, refers to simple mucous inflammation, or ulceration. I shall, therefore, proceed to describe the use of nitrate of silver, as the standard treatment (if I may be allowed such a term) of inflammation and ulceration of the mucous membrane of the os and cervix uteri. Before doing so, however, I wish to draw a broad and well-defined line between cases to which these stimulants and caustics are applicable, and those to which local depletion and counter-irritation are adapted, as the local means best suited to them. And in order to be understood, I will again draw the attention of the reader to the fact, that when a mucous membrane is inflamed, touching it gives to the sense of the patient the idea of rawness; when a part is touched in which the inflammation is beneath the mucous membrane, the idea of tenderness is experienced. When the mucous membrane of the cervix, for instance, is the exclusive seat of disease, if there is any disagreeable feeling experienced upon touching it, it is that of rawness; but if the substance of the cervix or body of the uterus is inflamed, when it is touched by the finger, or an instrument, the patient complains of tenderness. We should bear in mind, too, in estimating the value of the sense of tenderness in distinguishing between mucous and submucous inflammation, that we may sometimes be deceived by the complaints of patients, when the mucous membrane of the vagina is inflamed, into the opinion that inflammation is in the uterus. We ought, therefore, successively to press upon the different parts with our finger in a digital examination, and, after the speculum is introduced, with the probe, and question the patient, when each point is touched, as to the sensitiveness at that place. When pressing upon the uterus with the finger or probe, if the patient complains of tenderness or soreness, we ought to suspect submucous disease. Now, when the uterus is very slightly if at all tender to the touch, it is not likely that there is much submucous disease. To the mucous inflammation, these stimulants, astringents, and caustics are adapted, and to a more limited extent to the submucous. We very frequently find the increased secretion, the pus-colored mucus and rawness, combined with the deep tenderness and tense pain of submucous inflammation. In these cases we should be careful to subdue this last by depletory measures, alteratives, counter-irritants, &c., before we resort very freely to caustic and stimulant applications to the mucous membrane. When, however, there is evidence of inflammation of the mucous membrane of the cervix, outside or inside of the cavity of the body of the uterus only, a judicious employment of astringents and caustics will do more good for it than any other treatment with which I am acquainted. As this is the most numerous class of cases, and as separate submucous inflammation will come up for consideration after awhile, I will describe the treatment of it first, and the others afterward, premising that in mixed cases we should, to some extent, subdue submucous before we begin to use the treatment for the mucous inflammation; and in such cases, when we do begin to treat the mucous membrane with the caustics, we should do so with caution, lest we increase the deeper or submucous inflammation. I think this caution is not sufficiently understood, or acted upon. Too often the neck of the uterus is leeched, because it is inflamed, or it is touched with the nitrate of silver, because it is inflamed; and yet if the practitioner were to stop and think a moment, he would readily decide that leeching will not cure mucous inflammation, or that nitrate of silver is not applicable to submucous inflammation.

CHAPTER XII.

NITRATE OF SILVER AND ITS SUBSTITUTES.

CHRONIC inflammation is an habitual and established affection, having almost no tendency to spontaneous termination; it must be subverted to be cured. This can unquestionably be best done by local means, when the part affected is accessible. Inflammation of the cervix uteri is still less prone to spontaneous termination, from circumstances already mentioned, viz., the menstrual congestions, determinations of blood from its dependent position, and the excitement inseparable from the functions of the genital organs. On these accounts, the strong impression of nitrate of silver and of its substitutes is required. There can be no doubt but that the stronger the impression we can produce, the more completely the chronic inflammation is swallowed up by the acute, and hence the more radical the change; but if the impression is too strong, it may lead to greater damage than the disease for which it is used would produce. Doubtless, the white-hot iron which is recommended and used by some practitioners, causes more powerful effects upon the disease, more radically influences it than any application of nitrate of silver. But I think that we might not always be able to limit the extent of its influence within proper boundaries. The strong caustics are likewise more radical than the milder, and cure inflammation of the cervix more rapidly, and with as much or even greater certainty; but their effects are sometimes fearfully active, owing to an extension of the inflammation to other tissues than those to which they are applied. In order to avoid all likelihood of bad results from such extension of inflammation, the milder caustics are used, and their lack of power is compensated by the repetition of their use. As already intimated, the nitrate of silver is by far the most effective of these, in cases of inflammation and ulceration of the mucous tissue of the cervix. When the inflammation extends to the deeper tissues, it is not generally sufficient without the aid of other means.

The nitrate of silver is objected to by some as too strong and harsh a remedy to apply to so delicate an organ as the uterus, and speak of "burning the uterus" with lunar caustic as a "horrible operation." Honest observation, however, will convince every practitioner of intelligence that, with the precautions ordinarily enjoined, no more risk need be incurred by the use of nitrate of silver than by the use of any other valuable remedy. That there are cases to which it is not applicable, and in which it is too harsh, is certainly true; and it will be my endeavor to point these out, and enable the practitioner, by attention to the matter, to avoid damage from the nitrate in almost all cases. It is best that we should be aware of the fact that the nitrate is not infallible, nor always innocent; but we should also lay aside the unreasonable prejudice which arises from the term caustic, and which is hardly applicable to it, and determine, by our own observation, its title to the claim of a remedy in these cases.

Preparation for the Use of Nitrate of Silver.—All the preparation necessary, so far as the patient is concerned, will be effected in the examination for the purpose of clearly diagnosticating the disease; viz., the perfect exposition of the cervix uteri by the speculum, and the removal of all the mucus, blood, &c., by which it is often covered. This cleansing of the cervix from mucus, pus, or blood is important, from the consideration that these substances neutralize the effect of the nitrate by decomposing it.

Should be Pure.—In selecting our remedy, we should endeavor to procure a perfectly pure article, free from adulterations and impurities, as they act as diluents of it, and render the application less effective.

Form of Application.—It may be applied in the solid or fluid form. The former, I think, in the great majority of cases, preferable; while the latter, where the more concentrated solid form is too stimulant, may be made very useful.

Solid Form Best.—I am desirous of expressing a decided preference for the solid form, because its application may be made more easily, certainly, and definitely, and because the peculiar impression of this substance is thus more surely produced. The solid should be in the form of cylindrical pieces of half an inch in length; the size and form usually found in the shops. In some cases, the larger will be found most convenient, while in others we will use more easily the very small pieces.

Instruments for Using Nitrate.—I think a great deal depends upon the kind of instrument employed as a porte-caustique. In fact, we cannot expect to treat these patients successfully without having instruments that will expose the parts perfectly, and make the contact of our applications thorough and complete throughout. I am sure that many failures to cure arise from imperfection of instruments and want of thoroughness in application. It has been my lot, frequently, to be called to see patients of this kind in consultation with medical men who had been treating them for months with a glass cylinder for a speculum, and a goose-quill for a caustic-holder. In the very simplest of cases, where the inflammation or ulceration is all external to the os,—an uncommon thing,—it is only possible, it is certainly not probable, that success can be secured by such imperfect means of operating. To say the least of it, such treatment is

Fig. 28.

Small hard-rubber Syringe, to wash out the vagina or cleanse the neck of the uterus.

clownish. Let the practitioner have the best instruments to completely include in its exposure the whole of the cervix and the vaginal cul-de-sac; and to enable him to apply his remedy to all the inflamed surface outside the os, and inside the cervix, and, if need be, up to the fundus inside the corpus uteri. If, upon trial, his instruments do not enable him thus intelligently and thoroughly to proceed, he will do his patient and his own reputation injustice, as well as will misrepresent his profession, and will be utterly inexcusable if he does not invent, if need be, such means as will effect these objects. A porte-caustique for the solid nitrate, which I have used for several years, and with which I am very well satisfied, is made by Messrs. Tolle & Dagenhardt, instrument makers, in this city, and Tiemann, of New York. A large number of my medical friends have furnished themselves with this kind, believing it to be preferable to any of the common ones in use. The main feature is the flexible wire, of which a portion of it is made.

Flexible Caustic-holder.—It should consist of two pieces, one piece a sheath, about five or six inches long, and the other piece copper wire, about five inches long, surmounted at one end with platinum holders, into which the caustic may be fitted. These two pieces should be so made that when intended for use they can be screwed together, making an instrument ten or eleven inches long. When not used, the wire portion, holding the caustic, can be inserted into the sheath, thus making a caustic preserver as well as porte-caustique. We should be supplied with two or three sizes of these instruments, as a matter of convenience in cases where the os and cavity of the cervix differ in size. The object of having the stem made of copper or other flexible wire, is to enable us to bend it to suit the curvature of the uterus, or angle caused by a difference in the direction of the axis of the vagina and uterus. In many cases we cannot bring the cavity of the cervix and body of the uterus to correspond with the direction of the cavity of the vagina; in such instances a straight, inflexible porte-caustique will

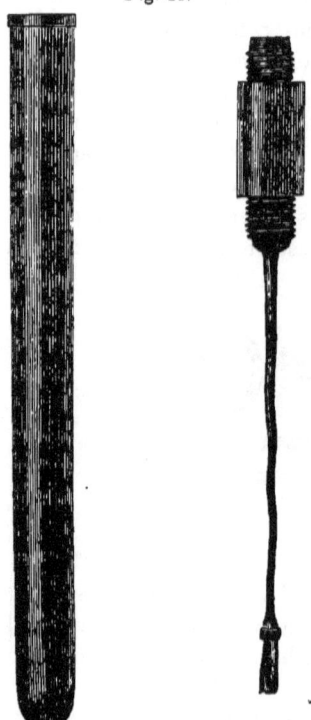

Fig. 29.

The flexible caustic-holder, in two pieces, to be screwed together when used, and sheathed by placing the wire part in the other when not used. (Full size.)

but very imperfectly enter and penetrate beyond the os; but if we have an instrument that will bend, and retain the flexure we produce in it, we may, as with the uterine sound, flex it so as to enter the cervix, and penetrate even to the fundus of the uterus. This flexible caustic-holder, or some other instrument that will answer the purpose of entering the uterus, I consider indispensable to success in a large number of the cases we are called upon to treat. The part to which the application is to be made should, in all cases, be divested of all mucus, pus, or secretion of any kind, before the medicine is placed in contact, and then it will act with more efficiency.

Mode of Applying It.—The nitrate should be applied thoroughly. Where there is inflammation external to the os, the nitrate should be deliberately and gently passed over the whole inflamed or ulcerated part; and should the disease extend inside of the os and cavity of the cervix, and even corpus uteri, it should be fearlessly but carefully carried up to the full extent of the disease. The contact should be perfect in every part, and sufficiently prolonged to produce all the effect it can produce by a single touch. If we use no more force than is necessary to keep the substance in contact with the part, there is no danger of keeping it there too long. This is the true "antiphlogistic touch," and it depends for this quality upon its completeness and thoroughness. Every time the application is made, we should try to be thus thorough in our use of it.

Frequency of Application.—This kind of application can be profitably made, as an ordinary practice, about once in six days; but we should be sure that all the perceptible influence of one application has subsided for at least twenty-four hours—and better if it is forty-eight hours—before another is made. This may require, in some instances, eight, or even ten days; or it may, in other cases, take place in five days. It is desirable, in making these applications, to avoid the period of menstrual excitement by not making it two days before the time for it, and waiting as much as two days after its complete subsidence. In most patients we will be able to make four applications a month; but often only three can be tolerated. In common ulceration or mucous inflammation, external or internal to the os uteri, we may expect to be under the necessity of making twelve or fourteen applications of this sort. In many cases more applications will be necessary, and in very few cases a less number will be required. Practitioners speak of curing their patients with three or four, some even with one application; but I am sure that they are nearly always deceived. Out of the large number of patients I have treated for inflammation and ulceration of the cervix, I have never known one to be cured with less than nine or ten applications. To the inexperienced I wish to say emphatically, be thorough in your applications, and be careful not to stop making them until every vestige of inflammation is gone.

Thoroughness and Perseverance in its Use.—Failures occur very frequently on account of too little being done by the caustic. Improvement is not a cure; nor are we warranted in believing that a patient, because she is better, will continue to improve until she gets well. The treatment must be persevered in until the cure is complete. I have observed, also, that regularity is important in the treatment of these cases. It will not do to visit the patient at our convenience; but we should see her, and make the application at the regular time, and attend to it promptly. It is not unusual, I think, for physicians to see their patients with so much irregularity as to fail in procuring the benefit of each successive application, and, to some extent at least, lose the advantage of one application before another is made. As I have already pointed out, a large number of cases are attended with inflammation in the cavity of the cervix; and in many instances, when there is no inflammation external to the os, the cavity of the cervix is the seat of much disease. We should remember this, and watch for it. I do not think it will be time wholly lost if I call attention more particularly to the mode of using the nitrate in these cases. As I have before stated, the continued discharge of pus, muco-pus, or even mucus to a considerable extent, is evidence of endocervicitis, and we should not cease treating these cases until this discharge has completely ceased. An entire cessation of the discharge from the cervix should be verified by the use of the speculum.

Application in the Cervical Cavity.—When the inflammation is in or extends to the cavity of the cervix, a flexible caustic-holder is indispensable to its successful treatment. We can be sure of making a thorough application inside the cervix—after exposing the neck in the speculum as fully and carefully as possible—by introducing the uterine sound into and through the cervix in order to exactly measure the direction and amount of curvature, and then bending the wire of the porte-caustique so as to correspond with the curve of the sound which has passed into the uterus. After this preparation, if the caustic-holder is not too large, it will readily pass into the parts surveyed by the sound, and thus bring the nitrate in contact with the diseased surface very completely, which, if allowed to remain in contact for a few seconds, will produce its full effect upon it.

Fig. 30.

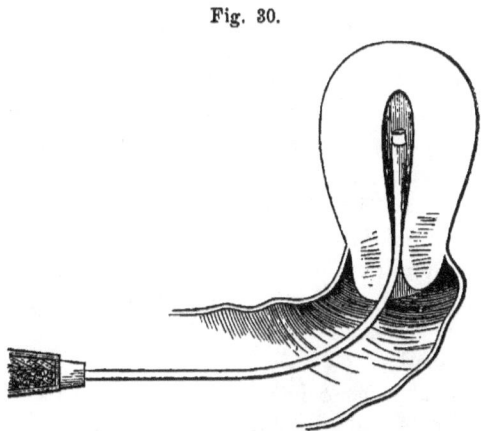

Showing the small-sized flexible Caustic-holder as it is introduced into the cavity of the uterus for endocervicitis and endometritis.

Solution, Strength, and Mode of Using.—The nitrate of silver is applied also in solution, and very often a cure is effected by it in this way. To be efficacious, the aqueous solution of the nitrate should be very strong; say one part of the silver to four parts of water. This solution is less powerful in its influence on the parts to which it is applied than the solid.

Frequency of using Solution.—It should be made consequently more frequently, every fourth day, for instance. Every part of the diseased membrane should be touched by it. We should not expect it to run upon the parts, but we should place it in contact by the instrument used.

A second edition of this work enables me to give the results of a more extended experience than I had when I wrote the first. I now have to say that I much more frequently resort to the solution of the nitrate of silver now than formerly, and believe that in many cases it is more easily tolerated, and produces better results. I generally use the solution in the strength of ℥iv to f℥j of water, and make the application, ordinarily, often as once in four or five days. This solution, although pretty strong, causes less of the inconveniences, such as pain and hemorrhage. I give a drawing of this simple little whalebone instrument. It seems in the drawing rather larger than the one I use. Unless very flexible and small, it will not answer the purpose completely. Almost all the instruments used for applications to the inside of the uterus are too large, and consequently the introduction is imperfect. This fact will account for many failures to cure where the disease extends deeply into the cavities.

Acid Solution of the Nitrate of Silver.—Another solution of the nitrate may be made by dissolving it to saturation in pure nitric acid. This solution is of course, very different to the other, and possesses the qualities of a powerful caustic. It must, therefore, be used with great caution, and subject to the rules for the government of the use of the acid nitrate of mercury, or other strong fluid caustics.

Instruments for using Solution.—The instruments I have used for applying medicines in fluid form to these parts, are the camel's-hair pencil, and a small swab, made by wrapping and fastening with thread a little cotton to the end of a fine piece of flexible whalebone. Either of these instruments will pass into the os uteri, if necessary, and carry along with it the solution. They may, of course, be made to apply the fluid to the outside of the uterine mouth with equal efficiency. The watery solution may be used profusely, as there is but little danger from contact with the sound parts. The acid solution must, on the contrary, be used very sparingly. The treatment of ulceration with the aqueous solution of the nitrate will require a long time, comparatively, for a cure; certainly twenty applications will be almost always necessary. But we should not stop the use of it until the ulceration, congestion, and hypersecretion all disappear.

I have introduced a figure to show the whalebone Fig. 31. with cotton on the end, to show the mode of preparing it, but it is larger than desirable. It should be small enough to be quite flexible, and easily enter and pass through the cervical cavity.

Effects of the Solid.—When the nitrate of silver is applied in the solid form to an ulcerated or inflamed cervix uteri, the first effect is to coagulate the albuminous compounds on its surface into a thin, very white film of the thickness of white writing paper. This film adheres to, and protects the surface from further influences that are not sufficient to destroy it; hence, after this effect, the nitrate can produce no more impression upon it. If, however, sufficient rudeness or force is used to separate this pellicle or film, from its adhesions, the nitrate will produce a similar effect upon the denuded surface, so that by forcibly passing the nitrate of silver over the mucous membrane for a number of times, a considerable amount of surface and substance, at some depth, may be destroyed. Taking these examples of its action, we see that it may be made to have a gentle or powerful caustic effect; hence the dispute as to whether it is or is not a caustic. If the nitrate is applied to the surface of a healthy mucous membrane, it produces the same effects, but more slowly and to a less marked degree. The difference in rapidity with which this coagulum or film is produced on the surface of an ulcer has been seized upon by Dr. Bennett as diagnostic of ulceration. On the ulcerated surface it is almost immediate, while on the membrane retaining its integrity, the film is formed more slowly. The surface of an ulcer becomes immediately white, that of the membrane slowly so.

Modus Operandi of Cure.—It has seemed to me that the application of the solid nitrate operates favorably, by two effects it has upon the diseased surface: the first is the profound stimulant action upon the capillaries, brought under its influence; and secondly, the formation of the film, which protects it from all foreign influence while it lasts. When this film falls off, it leaves the surface of the ulcer raw and bleeding, or if it has been applied to the mucous membrane, it is deprived of its epithelium. This occurs about the end of the third day, sometimes sooner, and sometimes later.

Discharge.—In the case of the ulcer, there is, after the loss of the film, quite a discharge of bloody serum, which lasts for forty-eight hours, or more. When this discharge ceases, it is on account of the generation of a temporary or permanent epithelium. Now, if the ulcer is examined, its edges will be found more defined, while its general surface shows an improved state of the granulations. After each time the discharge becomes less, the diseased surface smaller, until a completely healthy appearance is assumed. The application usually produces some pain, which lasts ordinarily from three to twelve hours. For an application to be beneficial, the pain should not continue longer than twenty-four hours.

Kind of Pain produced by Application.—The kind of pain produced by the application is not always the same. In simple mucous inflammation, it is apt to be of a burning or smarting character, or it sometimes merely increases the pains felt before the nitrate was used; the backache, pain in the side, or any other pain which had before existed, is increased, or, as the patient often expresses it, she feels the pain and other sensations which indicate the appearance of the menstrual discharge. In very many instances, the patient complains of no additional suffering. If the inflammation extends to the submucous tissue, the pain is apt to be more severe; it is soreness, a tense pain, or throbbing pain, and does not subside as readily as when the inflammation is confined to the mucous membrane.

Pain worse in Endocervicitis.—If the inflammation extends inside the cervix, and the nitrate is introduced into its cavity, the pain is apt to be somewhat more severe. Although all the local troubles are generally increased for a few hours, we meet with a few patients who are immediately, and very much relieved at the time of every application. This difference of suffering after each application is like what we observe with reference to the increase or decrease of symptoms after the beginning and continuance of local treatment. Some patients will suffer more after the commencement of local treatment for several weeks, and then gradually improve, and get well, while others will go on to convalescence from the beginning. Others, again, will not improve until the local treatment is finished. In addition to the increase of local symptoms under the application of nitrate of silver, patients are often rendered very much worse in their general symptoms. They are more nervous, their headache is increased, nausea is caused or increased; in fact, all the general symptoms enumerated as being caused by uterine inflammation will be found sometimes to arise from the effect of an application.

On the other hand, very frequently the general symptoms may be, and are permanently relieved by the local application from the beginning. If we observe through the speculum somewhat closely the effects of the application of the nitrate, we will find in the first place, and almost immediately, the ulcerated surface turn very white from the formation of a film of coagulated albumen. A short time afterward, the mucous membrane of the vagina and neck of the uterus become livid from congestion. In two days, or less time, the albuminous pellicle begins to be detached, and the surface beneath is left of a scarlet red, and often blood may be seen exuding from this raw, uncovered surface. This exfoliation, or detachment, goes on for two days, until all the surface covered with the coagulum is left raw and bleeding; on the fourth or fifth day, this surface is again covered with a very thin epithelium, and the membrane ceases to bleed. The injected condition of the uterus and vagina, with the finishing of these processes, subsides. In four, five, or six days, the effects are all gone, and the capillaries begin to return to their old inactive state. Astringent injections do good by expediting all these processes, I think, particularly the subsidence of congestion of the vagina.

Chronic or Ultimate Effect of the Nitrate upon the Tissues.—The chronic effects of nitrate of silver—by which I mean the permanent influence it produces upon the tissues of the uterus—are worth closer study, and I should be glad to give them to some extent; but I propose at present, for want of time and space, to confine myself to a very limited view of them.

Atrophy.—Sometimes the continued application of nitrate to the mucous membrane of the uterus induces condensation of the tissues beneath it, as well as in the mucous membrane itself; hence results, not unfrequently, true atrophy of that organ.

Contraction of the Os.—In very many cases, where the application is made to the os uteri for several months, that orifice becomes so small as to be of the size of a mere pinhole. This may sometimes take place while there is still inflammation in the cavity of the cervix. When this is the case, the secretions issuing from it will sufficiently indicate it. We need not be embarrassed in our treatment by this occurrence, as we can easily dilate the os uteri to almost any extent by tents of compressed sponge, or, if this is not at hand, slippery elm bark bougies. By using one of these tents, or bougies, for twenty-four hours before we desire to make the application, the opening will be large enough to answer all purposes. This contracted condition of the os uteri, where there has or has not been treatment, should not deceive us in reference to the presence of inflammation in the cervix. I have not unfrequently been called to see cases in which the mouth of the womb was scarcely perceptible to the eye on account of its contraction from inflammation or the use of the nitrate; several of which had been pronounced to be in an entirely healthy condition. Yet, from this minute opening, quite a large amount of muco-pus or tenacious mucus found its way in the twenty-four hours, and could be seen filling up the upper part of the vagina. By dilating the os uteri with sponge or slippery elm, and applying nitrate of silver inside the cervix for a number of times, all the distressing symptoms and the copious secretion subsided together.

Effect upon Menstruation.—The menses are ordinarily rendered more easy and natural by the cure of the inflammation from the use of the nitrate applications; but this is not always the case. At first, the sanguineous flow is increased; this may be for the first, and even second month, but in some instances, after this; it then diminishes to a great extent, so as to amount almost to amenorrhœa. I think this diminution of the menstrual flow keeps pace with and is dependent upon the condensing or atrophizing of the tissues of the organ. I have noticed this to occur so often that I regard it as a sufficient indication for the withholding of this remedy altogether when this condition is observed. This atrophizing and amenorrhizing influence of the nitrate is much more apparent after its introduction into the cervix and uterine cavity. I do not remember to have seen atrophy result from treatment with any of the substitutes for the nitrate. Sometimes the menstrual diminution results apparently from the effects of the frequent application of the nitrate to the mucous membrane of the cavity of the cervix and corpus uteri; while, so far as we can judge from examination, there is no diminution in the size of the uterus, nor, where it seems to be hardened in consistence. When this is the case, it is doubtless on account of the transforming influence exerted upon the mucous membrane, perhaps a condensation of its structure to such an extent as to prevent the capillary fractures which in health allow the transudation of the menstrual blood.

Effect in Dysmenorrhœa.—Painful menstruation is modified to a greater or less extent by the application of the nitrate. For the first, and even second month, there may not be much difference, but after this the painfulness ordinarily diminishes until it ceases, or nearly so. Sometimes, however, at the first recurrence after the beginning of treatment, the pain is almost entirely relieved. This is remarkably the case in cases where the pain has been of a cramping instead of an aching or burning nature.

How are we to know when to stop its Use?—How can we know when the nitrate has been sufficiently used? We are to continue the treatment, as I have before said, until every vestige of inflammation is removed. We must continue the applications until all the ulceration is removed that is within our sight, and then continue them in the cavity of the cervix, and, if need be, the cavity of the body, until no free mucus is seen issuing from the cervix or in the vaginal cavity. It is a mistake to suppose that the inflammation is cured until the pus or mucus, or both, which are evidences of its existence, cease to appear when we make our examination. I cannot emphasize this direction sufficiently to do justice to its importance. While there is yellow or puriform mucus in any quantity issuing from the os uteri, there is ulcerated mucous membrane within the cavities above, which require the use of the applications; while there is hypersecretion or free mucus issuing from the os uteri persistently, there is inflammation or persistent congestion of the mucous membrane of the cavity of the cervix, which requires the lunar caustic for its cure. We should continue it, therefore, until these cease to flow, as well as until the obvious ulceration is healed.

The Nitrate sometimes fails.—The nitrate sometimes fails to cure these inflammations and ulcerations, and although it may not be considered necessary by the reader to inquire into the causes of its failures, yet I think we will treat these cases more successfully by rightly understanding why we do not succeed by the use of the ordinary remedies.

Not strong enough.—There are cases in which it falls short of producing the impression necessary to arouse the capillaries to a more healthy action; it is not sufficiently powerful. In these cases no apparent or real good is done, but the inflammation continues about the same all the time. The cases in which it fails in this way are generally indolent; the granulations are large and flabby, the cervix large and doughy to the touch, with very little sensibility, and the surface is inclined to bleed easily.

Substitute in such Cases.—These require some of the stronger substitutes, applied occasionally, and alternated with the nitrate, or with some of the milder substitutes. The caustic potassa is much the best substitute in such cases. My plan of applying it, in such cases, is to moisten a very small camel's-hair brush with the mucus of the vagina, and rub it over the stick of caustic potash until the brush becomes well saturated with it. I then apply the brush to the diseased part. I continue to apply the mucus solution of the potash to the surface in this way until the desired effect is produced. In this manner we may procure a strong stimulant influence, or slight caustic effect, without the destructive substances running upon the sound parts. A swab, made by tying a small piece of cotton to a small stick of whalebone, will answer the purpose equally well. We first moisten the cotton swab in the thick mucus, and pass over it the stick of caustic until it dissolves off and retains a part of it, and then apply it to the diseased part. Or we may dip the brush or swab in strong nitric acid, and apply to the parts. The swab I think the better of the two, as it does not take up the caustic fluid so freely, and hence is not likely to allow it to flow over the sound parts.

Sometimes the Nitrate fails without apparent Reason.—But we often meet with instances in which, without any apparent reason, the nitrate fails to do any good. These cases we should study, with a view to ascertain whether the impression is not sufficiently powerful, or whether the impression is not of the right sort, and select our substitute according to our conclusions in this respect.

May cease to do Good after being Beneficial.—Again, the nitrate may do good and seem to be curing a case, but, after several applications, there seems to be no advance. The ulceration remains the same from week to week, without any change. It will be necessary, in these cases, almost invariably to resort to some stronger stimulant, as the acid nitrate of mercury, the acid nitrate of silver, or the caustic potash, with the brush or swab.

Acid Nitrate of Mercury and of Silver.—The acid nitrate of mercury can be procured at the shops; the acid nitrate of silver is made by dissolving the nitrate of silver in the strongest nitric acid, to saturation. Any of these may be tried once a month, to be succeeded by milder substitutes, as tannin, sul. cupri, creosote, &c., at intervals of a week between them.

Sometimes the Nitrate of Silver does Harm.—But sometimes the nitrate of silver not only fails, but it entirely disagrees with the cases, and it has to be abandoned. I have known a number of cases in which the nitrate aggravated the inflammation every time it was applied.

In Aged Persons.—This is particularly apt to be the case in old persons, after the childbearing age has passed. In them the inflammation assumes, nearly always, a peculiar appearance; the cervix is small, the granulations minute, the surface very red, and the discharge a thin and acrid muco-pus. These are apt to be obstinate, and almost invariably made much worse by the application of the nitrate, and, what seems singular, are benefited by the stronger stimulus of potassa fusa. One application of caustic potash, with the brush or swab, every four weeks, followed every six days with tannin, usually answers very well for this kind of ulceration. Creosote generally agrees well with it.

Aphthous Inflammation.—Another sort of inflammation, attended with patches of exudation not unlike aphthæ, is almost invariably very much aggravated by the application of the nitrate of silver. This requires milder treatment. Tannin and creosote, alternated every six days, with one application, if necessary, of caustic potassa, will answer very well. On several occasions, I have found the nitrate, after having done well for several weeks, suddenly and unaccountably to disagree with cases, and the ulceration spread rapidly. These have been rendered tractable by the caustic potash, pretty freely applied.

Causes too much Discharge.—But without reference to the kind of ulceration, the nitrate of silver, so far as I am able to judge, sometimes disagrees and does harm, on account of the excessive discharge or hemorrhage it causes. Ordinarily, when the nitrate is thoroughly applied, as I have elsewhere said, there is some discharge of bloody serum, amounting to half an ounce, or double that quantity. This takes place from the second to the fourth days inclusive. Sometimes it is much less, sometimes it is more abundant. I have met with instances, however, where there was great hemorrhage, and was so exhausting as to preclude the use of the nitrate entirely. So far as I can see, there is no peculiar appearance by which we might be led to suspect the occurrence of hemorrhage, before trying the remedy. One case that I am now treating is peculiarly susceptible in this respect. A single application of nitrate of silver, in the middle of her menstrual month, caused her to flow so copiously as to make it necessary to keep her bed, use cold applications, and acid drinks. In spite of these, she lost fifteen or twenty ounces of blood in eight or ten days. This was repeated the next month, and it became necessary to abandon the remedy altogether.

This, of course, is a remarkable case, but in many instances so much loss of blood has taken place as to cause me entirely to forego its use in those cases. In the cases in which hemorrhage forbids the use of the nitrate, the substitutes I have found most suitable are the caustic potassa and tannin. The caustic potassa may be used once in the middle of each menstrual month, with the little cotton swab I have described, so as thoroughly to stimulate the inflamed part and produce very little cauterization; and every fourth or fifth day, in the intervals, completely saturate the inflamed surface with pulverized tannin, applied with the camel's-hair pencil or the swab. Before using the tannin, we should entirely remove the viscid mucus in the neck and about it. We need not be apprehensive of any severe effect from the tannin, either in the cavity of the cervix, or on its external surface; we should apply it fully and freely to the whole inflamed surface. Creosote, alternated with the tannin every fourth or fifth day, often suits such cases. When the ulceration is external and extensive, in these bleeding cases, it is best generally to apply the caustic potash in the solid form, so as to produce a more profound effect.

Nitrate sometimes causes too great Pain.—Too great pain is sometimes the result of application of nitrate of silver. The pain, after application of the nitrate, may be merely slight, the patient scarcely feeling any inconvenience whatever; or, what is usual, it may produce some pain and suffering in from six to twenty-four hours, and then subside; or, in rare, exceptional cases, cause intense pain.

The pain, when severe, may subside in a few hours, and is not worth making any change in the remedies, or the pain may be severe and protracted. When this last is the case, injurious instead of beneficial effects are the result, and we should seek for a substitute. Caustic potash, tannin, creosote, acid nitrate of mercury, or some other acknowledged substitute, should be employed. The acid nitrate of mercury is an excellent substitute in such cases, alternated with the tannin, &c.

Worse in Cases of Submucous Inflammation.—This local pain, after using the nitrate, is more common where there is some submucous inflammation; a few leeches will frequently remove the disposition entirely.

Without these local pains, or other suffering with them, there is, as the result of the application of the nitrate to the os and cervix uteri, sometimes excessive nervous symptoms. The nervous excitement sometimes becomes so great that it is very alarming. A patient upon whom I attended but a few months since, was rendered entirely sleepless, and almost insane, by the exciting influence of those applications, and it was necessary to send her off to the country for tranquillity and recuperation. In quite a number of instances which have come under my observation, the nervous symptoms were so increased, that I had to change the treatment, or use substitutes that would not produce these peculiar effects. It is singular, that these very nervous patients complain very little, if at all, of the local effects of the application, and are only rendered nervous by it.

I should hardly finish what I ought to say of the nitrate of silver, and its substitutes, were I not to state that the latter do not cause any of these symptoms of distress which I have mentioned as the occasional result of the application of the former. There is something, then, peculiar and distinctive in the influence of the nitrate of silver, as evidenced by this fact. It is not merely stimulant, astringent, or caustic, in its effects upon this inflammation, but it has its own peculiar influence. It may be asked if the nitrate causes these bad effects sometimes, and none of its substitutes ever do, why use the nitrate at all?

In the first place, when it does agree with a case, there is no remedy that acts so kindly, so efficiently and certainly, as this. In the second place, the weaker substitutes are slower and less certain than the nitrate, and consequently, when successful, take more time to make a cure. And in the third place, the stronger caustics, as the acid nitrate of mercury, the acid nitrate of silver, and the caustic potash, require greater care, and any accident occurring from them may be much more serious, and, if carelessly or awkwardly used, are likely to do damage to parts not intended to be influenced by them. The nitrate requires almost no preparation or precautionary measures for its use; for the stronger substitutes we must prepare carefully, and use much precautionary vigilance. The nitrate in solution does not produce such decided effects as the solid, and hence, of course, will not cause hemorrhage, pain, or nervousness to the same extent that the latter does. Can we continue to use the nitrate when it causes the above inconveniences, and counteract or neutralize its effects by some other remedy? The pain and hemorrhage are apt to become less at each successive application, and hence, if the patient can bear them for a few times, we may continue to employ them, and then the cases are generally cured by them; but occasionally they disagree after having acted kindly for a time.

Remedy for the Hemorrhage.—When the hemorrhage is considerable, Dr. Bennett recommends a plan which I have followed with good results sometimes, and that is, to make the application to only a part of the ulcerated or inflamed surface. When the application is extended inside the cavity of the cervix, this direction cannot be observed. And it is in these cases that the hemorrhage is the worst. Astringent injections and cold applications, baths, &c., when the hemorrhage is not very great, will afford some relief and enable us to go on in their use. Generally, however, we will have to do with the substitutes when the hemorrhage is considerable.

Remedies for the Pain.—When the pain is great, emollient injections of linseed infusion, infusion of slippery elm bark, with laudanum, thrown into the vagina in large quantities, or half a teaspoonful of laudanum in a little starch-water or linseed tea, per rectum, will also aid very much in quieting. It is better in all cases for the patient to remain still in the recumbent position, for some hours after an application; when there is much pain, it is indispensable. The patient should be quiet until the pain is over.

Remedies for Nervousness.—When the nervous symptoms are excessive, we should be cautious about repeating the applications. If opium does not disagree with patients, its anodyne influence may enable us to continue the treatment. Tinct. hyoscyamus and camphor may also be tried, or valerian, brandy, &c. But some of these, particularly the last, must be used sparingly. If the nervousness subsides in a few hours, either with or without the aid of the anodynes, we can still resort to the nitrate applications. But if it continue at all obstinate, we must use some of the substitutes. I can but remark again, that it is singular that the caustic potash, and all the stronger caustics, produce less pain, less hemorrhage, and less nervous excitement, than the nitrate of silver.

Is its Application allowable in Pregnancy?—Is it ever allowable to apply the nitrate to the cervix uteri, inside or out, after the commencement of pregnancy? I confess that I am afraid to do so, and if a patient becomes pregnant during treatment, I advise a discontinuance until after confinement, and complete involution has taken place; say three months after accouchement. I know that Drs. Bennett and Whitehead both advise the use of the nitrate during the first three months, for the purpose of avoiding abortion, but the great irritation it sometimes causes intimidates me from using it, or recommending others to do so. I think I have seen abortion caused by it, in cases where pregnancy was not suspected. On the other hand, I have seen cases where pregnancy was not thought to exist, treated for some time without any bad effects. Upon the whole, I think it is much the best practice to desist after conception, or not to begin if we know it has taken place.

Loss of a Piece in the Cervix.—Some object to the introduction of the nitrate of silver, in the solid form, into the cervix uteri, lest a piece of it accidentally be left in that cavity, and very bad results follow. I have had this accident to occur to me repeatedly, and as yet I have not seen any bad results from it. It is true, the pain is sometimes a little more severe and protracted in duration, but it dissolves and runs out, or is expelled into the vagina, which is the more probable course, and there is dissolved and neutralized by the mucus of that cavity. I have been so strongly impressed with the harmlessness of the presence of a small piece of the nitrate there, that I have, in certain cases, intentionally passed some up the cervix, and allowed it to dissolve in the fluid and distribute itself over the surface of that cavity.

Injecting Ointments.—In almost all cases of inflammation of the cervix, the disease extends through the cavity to the cavity of the corpus uteri. The difficulty of making a speedy and perfect cure is often owing to this fact, and I think more of the failures to cure chronic inflammation of the uterus are attributable to the want of efficient methods of reaching that intra-uterine disease with the proper remedies than any one other cause; hence the great importance of being well supplied with resources for this purpose. I have already expressed my decided disapprobation to injections in the cavity of the uterus. A process for the introduction of remedies in the cervical and uterine cavities, in some respects similar to injections, is the deposition of ointments containing medicines that are desirable, by means of a syringe made with that intention. There is no question but that many obstinate cases of endometritis may be thus reached and cured. I should strongly urge the practitioner to be cautious not to introduce a large quantity, and especially in the cavity of the body, as I have, in two instances, been greatly alarmed at the acute inflammation thus caused, although both patients did well in the end.

The ointments may contain nitrate of silver, ten grains to the ounce of simple cerate, or ℥ij of ung. hyd. nit. to same quantity, or tannic acid, acetate of lead, &c.

I give a wood-cut of an instrument well adapted to such use. The ointment is placed in the barrel of the syringe, the tube inserted into the os uteri, and carried as far up as the practitioner considers necessary, where the deposit may be made by moving

Fig. 32.

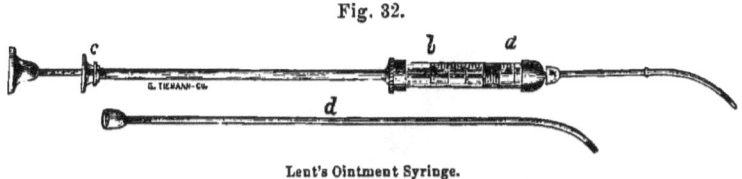

Lent's Ointment Syringe.

the piston forward. These applications may be repeated once in five or six days.

Tents or Bougies.—Still another method, which I have practiced with more success, perhaps, than any other, of making a curative impression upon the internal membrane of the uterus, is accomplished by introducing tents or bougies, either medicated or not. I use them much more frequently without medicating them, and believe that, unless in special instances, they are better thus employed. The slippery elm bark is the material of which I make them generally. My students make them for me in considerable quantities, using their pocket-knives as the only instruments necessary. I give a cut of one of the ordinary size and shape. A very important thing in making them is to place the thread about them very securely, as they sometimes pass entirely within the cavity of the uterus. When this accident occurs, and they cannot be removed on account of the loss of the thread, they give the patient considerable pain, and cause a profuse discharge of muco-pus, but gradually melt away, and in ten or fifteen days are gone. I have met with this accident several times, and have not given myself the trouble to dilate and extract, as it soon comes away itself. In order to their introduction, I expose the os uteri as for any other application, moisten the tent, flex it with my thumb and fingers, and assure myself that it is perfectly soft and pliable, so that it may make any curve necessary without violence or even resistance, and then with the dressing-forceps pass it up entirely inside the cavity. If the end of the tent is passed entirely within the

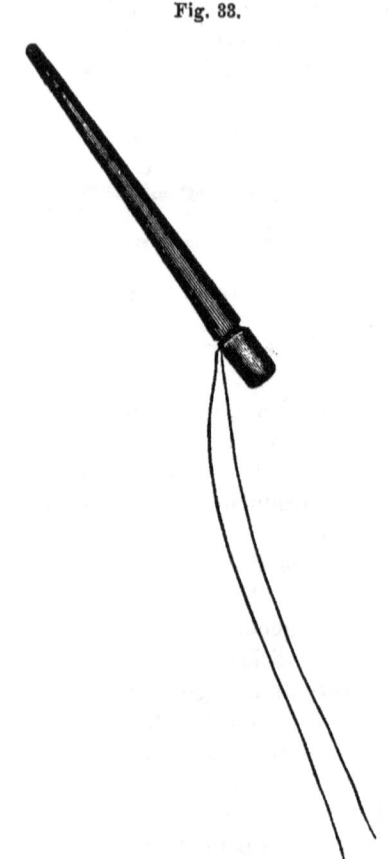

Fig. 33.

Slippery Elm Tent.

cavity, so that it disappears, it will remain until removed by traction upon the thread, which should be long enough to hang out of the vagina and be easily found. After it is introduced, I say to my patient that if it causes much pain, to remove it in two hours, but if it does not give great pain, to allow it to remain for twelve hours, and then draw upon the thread until it comes away. I often use the bougies in this way, almost to the exclusion of other intra-uterine applications, in certain cases, with gratifying success. Indeed, I regard them as among the most

Fig. 34.

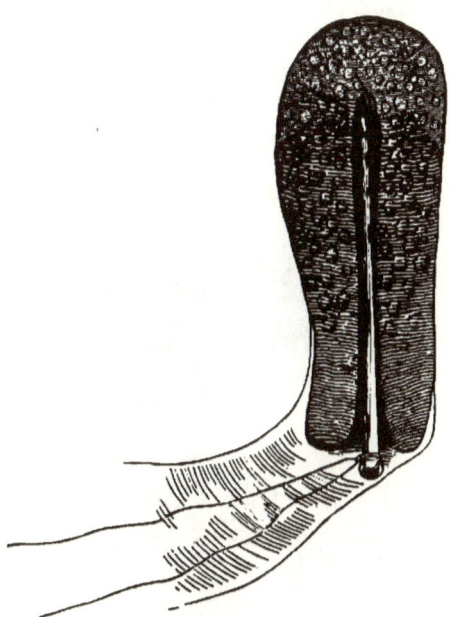

Slippery Elm Tent introduced.

valuable means of stimulating to altered action of the mucous membrane of the cervix and body. They may be dipped in the solution of the nitrate of silver, tincture of iron, creosote, tannic acid, or other substance, and then introduced in the same way. When I first began to use them, I supposed it necessary to have them large enough to cause pressure upon the mucous membrane, but further observation has convinced me that it is not necessary to have them large enough to fill the cavity. Remaining in the cavity as a foreign body, they cause an alterative afflux of blood to the diseased part, not dissimilar, perhaps, to the stimulus of a medicinal application.

The slippery elm tent is especially valuable in cases where there is inflammation in connection with a sharp curve. In such cases the caustic-holder or the tube of the ointment syringe cannot be flexed sufficiently to pass the curve, while a small, soft, moist slippery elm bougie will pass very readily, and answer the double purpose of opening the canal so that fluids will easily pass, and cure the inflammation. The practitioner who will give them a fair trial in this class of cases, will be gratified with the result. The dysmenorrhœa generally attendant, and so distressing in this double difficulty, flexure and inflammation, is very generally at once benefited by it, and finally cured. This use

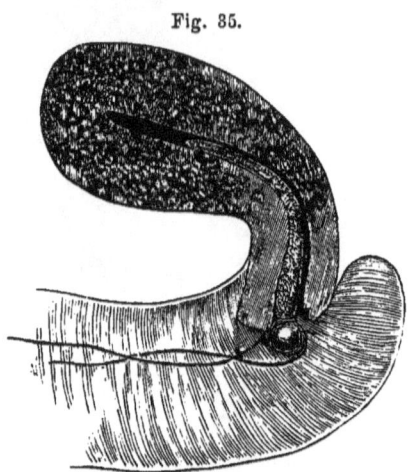

Fig. 85.

The Uterus in a state of anteflexion, with the Slippery Elm Bougie introduced into it.

of the slippery elm tents enables me now to dispense with incisions for the rectification of the canal, or for the dilatation of a narrow cervix. And it is one of the very best means of overcoming the great suffering caused by flexures. The use of the tents must be followed up for a sufficient length of time to assure permanent success. I have no doubt but that India-rubber tents could be made that would be a fair substitute for this material, but it is so easily obtained that I have not tried to supply myself with that sort of substance.

Sponge tents, with and without medication, are also very profitably employed for the cure of endometritis, and especially endocervicitis. The cases to which they are applicable are such as are attended with a more patulous os uteri, and a flabby and hemorrhagic condition of the mucous membrane of the cervical cavity. They should be introduced well up, and allowed to remain ten or twelve hours; and they may also be withdrawn by the patient if they are secured by a strong piece of twine. The inexperienced should be made aware that the size of the tent should be small enough to pass in without much force, as they soon distend sufficiently to cause pressure and vascular excitement. They, I think, should always be impregnated with carbolic acid or other disinfectant fluid, to prevent the blood and mucus from undergoing decomposition, and thus causing toxæmia, which I have seen result from their use. In the class of cases here indicated they will be very valuable. We may resort to them as often as once a week, or once in two weeks, alternating them with other applications if desirable to do so.

I subjoin a summary of the treatment of Robert Ellis, Esq., Obstetric Surgeon to the Chelsea and Belgrave Dispensary, London Lancet for July, 1862, reprint. It is a choice tabular view of the kinds of ulceration we meet with, and the very best mode of treating them, and I think will be useful to the inexperienced:

VARIETY.	TREATMENT.
1. *Indolent Ulcer.*—Cervix hypertrophied, of a pale pink color, and hard. Os patulous to a small extent. Ulcer of a rose red. Granulations large, flat, insensitive, and edge of the ulcer sharply defined. Discharge: mucus, with a little pus, and occasionally a drop of blood.	For a few times the caustic pencil—solid nitrate silver. Afterward, the solution of nitrate of silver in strong nitric acid.
2. *Inflamed Ulcer.*—Cervix tender, hard, a little hypertrophied, hot and red. Vagina hot and tender. Ulcer of a vivid red. Granulations small and bleeding. A livid red border around the ulcer. Discharge: a muco-pus, yellow and viscid, with frequently a drop of bright-red blood entangled in it.	Occasional leeching, hip bath (warm), emollient injections. Then acid nitrate of mercury several times, succeeded by the solar lunar caustic, potassa fusa, or cum calce.

VARIETY.	TREATMENT.
3. *Fungous Ulcer.*—Cervix soft, large, spongy to the touch. Os wide open, so as to admit the finger. Ulcer large, pale, studded with large and friable granulations. Discharge: glairy, brownish mucus, frequently deeply tinged with blood.	At first, the caustic pencil. Subsequently, nitric acid, solution of nitrate of silver, or acid nitrate of mercury; electric, or actual cautery.
4. *Senile Ulcer.*—Cervix small, red, a little hard. Ulcer small, extremely sensitive, of a bright-red color. Granulations very small, red, and irritable. Discharge: a thin muco-pus.	Potassa fusa, or strong nitric acid, with nitrate of silver once or twice at long intervals. Then solid sulphate of copper, in pencil.
5. *Diphtheritic Ulcer.*—Cervix of ordinary size, a little hot, dry, and tender. Ulcer covered in patches with a white membrane, adhering closely, irritable, and readily bleeding beneath. Discharge: a thin acrid mucus, without pus, but occasionally tinged with blood.	At first, electric cautery, potassa cum calce, or acid nitrate of mercury, two or three times at long intervals. *No nitrate of silver.* Subsequently, stimulant applications, tincture of iodine or sulphate of copper.

CHAPTER XIII.

TREATMENT OF SUBMUCOUS INFLAMMATION.

Submucous inflammation, as has been seen, is observed under a variety of accompanying circumstances; with mucous inflammation, without mucous inflammation, and without change of size or consistency, and with fibrinous deposit, enlargement, and induration of the cervix. Of course, all these circumstances will more or less modify the treatment of the different cases in which they are observed to occur.

Submucous Inflammation, with Ulceration and Mucous Inflammation.—There is often evidence of submucous inflammation when ulceration affects the mucous surface of the cervix. When the tenderness is not considerable, nor the part enlarged and tumefied, the cure of the mucous disease by the means heretofore indicated, will suffice to cure the submucous also, and hence the case will need no further treatment whatever. But if the cervix is quite tender to the touch, somewhat swollen and hot, and the ulcerated surface red and excavated, and giving out pus copiously, other remedies than those adapted to the cure of the mucous inflammation ought to be used. Leeches, in numbers to suit the intensity of the inflammation and the general condition of the patient, should be applied, and repeated every week, until the tenderness and heat have subsided to a great extent. But as local depletion will not always produce the effect, it will almost always be better to resort to internal alteratives and sedatives. Very many of these cases yield promptly to the alterative influence of mercury, gradually induced, with an occasional active cathartic of salines. Probably the best general plan is to leech the cervix, give a cathartic of calomel, to be rendered a little more active by sulph. magnesia, citrate of magnesia, Seidlitz powders, or Congress water. If, after two or three days, the local tenderness, pain, and heat continue, it will be well to give a grain of proto-iodide of mercury, or calomel, in similar doses, combined with opium, every four or six hours until slight ptyalism is produced. The cathartic, depletory, and alterative treatment should be continued until the submucous portion of the disease is removed; when the inflammation of the mucous membrane may be treated as I have directed when particularly speaking of that subject. As leeches are not always attainable, it becomes a matter of interest to find a substitute for them. We have this, fortunately, in scarifications; a remedy to which we may resort without apprehension when local depletion seems to be indicated.

Fig. 36.

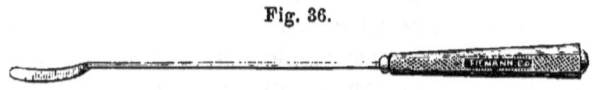

Scarification—Mode of doing.—The mode of doing this is practiced differently by different individuals. The plan which I have found most convenient and effective is to make the incisions in the os uteri, and direct them somewhat outward. A knife with a probe point is best adapted for this operation. The blade should be about two inches long, and one-eighth of an inch wide, very thin, and mounted upon a light straight handle, seven or eight inches long. After the neck and mouth of the uterus are brought full into view by the speculum, the probe point may be introduced half an inch through the os into the cavity of the cervix; when thus placed, the handle should be carried as far to the side of the speculum to which the edge is directed as allowable, and then withdrawn with enough pressure to make the edge cut through the mucous membrane. This will allow of a considerable flow of blood. If we wish to obtain

Fig. 37.

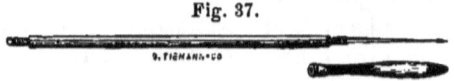

Dr. Buttle's Uterine Scarificator and Leech, very efficient and convenient for abstracting blood from the engorged cervix uteri.

a large amount of blood, several of these little incisions may be made around the circle of the os. The copiousness of the flow may be regulated by the depth as well as number of these incisions. In a few days ordinarily all trace of these little wounds is lost, and the os resumes its usual appearance. This is the class of cases to which the depletion so often directed is very well adapted. The indication for the depletion is in the tenderness, heat, and swelling, all of which are dependent upon submucous inflammation, and not upon the ulceration, or other signs of mucous inflammation. Depletion has, indeed, but little if any good effect upon inflammation of the mucous membrane.

Leeching.—Although it may seem hardly necessary to give any direction with regard to the mode of applying leeches to the cervix, it may be useful to the young practitioner in the treatment of these cases to do so. A common glass speculum introduced so as to include and isolate the cervix, is what I have been in the habit of using. The leeches are thrown down to the bottom of the speculum after the parts are cleaned of mucus, and the leeches are watched until they seize the part. Three are about as many as can conveniently be used in the speculum at one time. If more are considered necessary, they must be applied after the bleeding from the bites of the first has pretty well subsided. The bleeding usually continues for several hours, and as much or more blood is lost, after they are removed, as they draw. Submucous inflammation sometimes outlasts the mucous, and when we have this last cured, the troublesome symptoms still continue, or very seldom the submucous begins and continues independent of the mucous part of the disease. These cases, unattended by hypertrophy or induration, as they sometimes are, cannot be diagnosticated by the speculum alone. The physician is rather surprised, perhaps, that the symptoms of uterine inflammation should continue after an examination with the speculum shows a perfectly healthy color, size, and secretion of the organ; yet this is sometimes the case. If the sound or finger be pressed against the parts, they will be found to be tender. This condition is not unfrequently left after the cure of chronic mucous inflammation, and keeps up the symptoms. When it is a sequel to the chronic mixed form of mucous and submucous inflammation, it is apt to subside spontaneously after a time. When it exists independent of mucous inflammation, and is not the sequel to the mixed form, it is often, though I think, not necessarily, connected with scanty menstruation. Not unfrequently in this variety of the disease, the uterus is smaller than natural. It is quite common, when attended with scanty menstruation, to attribute it to this last circumstance; but I am inclined to think the deficient menstruation and atrophy are both attributable to the inflammation as a cause. It would be irrational to stimulate the uterus to greater congestion, to increase the flow when it was thus the subject of inflammation. The treatment should be directed to the inflammation. The remedies used will depend upon the acuteness of the symptoms; if the pain is considerable, or the tenderness great, leeching moderately or cupping upon the sacrum is quite desirable, but it should be used only to remove the tenderness and pain; the subsequent treatment should consist in the use of alteratives, counter-irritants, and tonics. Small doses of mercury, followed by the saline purgatives, as alteratives, answer admirably; six grains of blue mass every fourth or fifth night, followed in the morning with Epsom salt, seidlitz powders, or citrate magnesia, is a good alterative. One grain of calomel may be substituted for the blue mass, if the patient is plethoric. If the patient is anæmic or weak, the bleeding should not be resorted to, but the alterative may be accompanied with tonic treatment. The preparations of iron, syrup of the iodide, syrup pyrophosphate, iron by hydrogen, or the tincture, are good and eligible tonics. I am very partial to the chl. tinct., given in twenty-drop doses three times a day. I have seen a great deal of good done by it in removing that congestive sort of inflammation that so often keeps up the sensitiveness of the organs, after the more active symptoms had been removed, or in anæmic and weak patients. When there is not much acuteness of inflammation, or necessity for depletion, much good will result from proper counter-irritation.

Seton as a Counter-irritant.—The seton is one of the best means for this purpose. It should be introduced and allowed to remain for several months, and caused to discharge pretty freely for the most of that time by occasional turning, and if necessary impregnation with some irritating powder, as cantharides, or savin root. It should be made of one whole, large skein of silk, or even larger, so that the impression may be powerful. The best place for it is over the symphysis pubis, or in cases where one of the iliac regions is the seat of pain, this is a desirable locality. I have sometimes directed my patients to dress the seton daily with mercurial ointment until gentle mercurialization occurred, with much resulting benefit, I have thought. Soothing injections often diminish the sensitiveness of the uterus; and if they do no other good, should be used for this purpose. Two teaspoonfuls of laudanum to a pint of water, to which thirty grains of acet. plumbi are added, make a very good injection. This should be passed into and through the vagina for several minutes. Belladonna, hyoscyamus, aconite, gelseminum, and cicuta may be used also with the same view, in proper quantities.

It is not a difficult thing with Dr. Skene's articulated uterine sound and knife to scarify the cavity of the uterus, and in certain cases of endometritis will do much good.

The figure here given will at once acquaint the reader with the instrument and its mode of operating. The blade is concealed when introduced, and then it is brought out by the nut in the scale and easily secured and measured at the proper place.

Fig. 38.

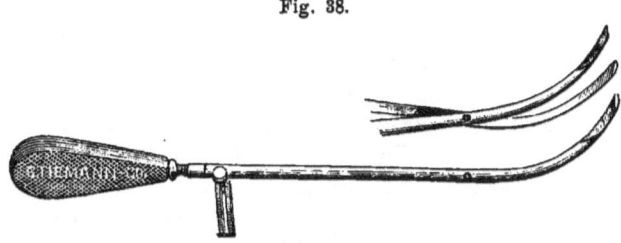

Hardness and Enlargement—Treatment of.—After the inflammation in the substance of the cervix has continued for a great length of time, fibrinous deposit hardens the tissue, and makes an enlargement which becomes permanent and difficult of cure. This enlargement and hardness are attended with various degrees of tenderness; sometimes the parts are not very sensitive to the touch, while in other cases the least touch causes exquisite pain and suffering. These conditions, of course, will very much modify the treatment. When there is tenderness, heat, and other signs of an acute condition of inflammation, leeching or cupping, cathartics, alteratives, anodynes, should be resorted to until these symptoms are removed or very much relieved. I have seen a slight ptyalism do away with these symptoms very quickly, and induce such a state of comparative comfort, that the patients believed themselves cured.

Treatment of Induration.—When the case is chronic, the cervix enlarged and indurated from the infiltration of fibrin, antiphlogistic means will not suffice. It will be necessary for us to resort to measures that will cause the absorption of the fibrin. The method of cure to be instituted requires that we use means that will cause congestion of the parts affected, and thus restore the vascularity of the tissues concerned. The vascular system of the uterus is peculiar, inasmuch as it may be stimulated to extension by anything that will hypertrophize the organ, and that foreign bodies introduced into its cavity will effect this hypertrophy and vascularization. We are enabled to seize upon these circumstances to promote the removal of the fibrin in the cervix, and we may sometimes accomplish it with considerable readiness by the employment of sponge tents.

It is not merely the slight forms of enlargement and induration that may be thus cured, but the greatest deviations in this respect may often be corrected as readily as with the more painful and hazardous remedies.

The sponge, to be most useful, should be carefully prepared by being well compressed, impregnated with carbolic acid, and covered with tallow. The sponges we find in the shops are often almost worthless, and it is best not to rely upon them unless we know them to be good. A sponge should be compressed so as to be not more than one-fifth the diameter that it was before prepared. Perhaps the easiest way for physicians is to moisten the sponge with a solution of carbolic acid in gum arabic mucilage, and compress it while wet by wrapping it strongly with hempen twine, and drying it with the twine around it. After it is thoroughly dry the twine may be removed and the sponge smoothed with a file, and shaped to suit the purpose for which it is intended, and then covered with mutton tallow, if it is to be used in winter, or spermaceti, if summer. In preparing it there should be a cavity in the large end sufficient in size to receive the end of the uterine probe or a staff made for the purpose, with which to introduce it. The sponges should be made of various sizes and lengths, and conical in shape. The patient

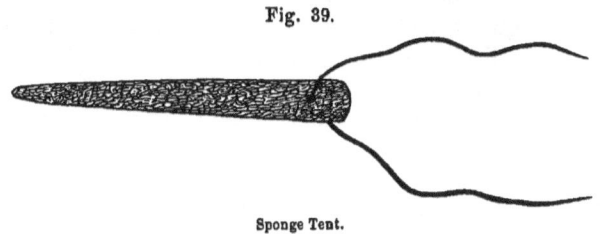

Fig. 39.

Sponge Tent.

will need no special preparation. A sponge of small size ought to be selected, one that may be passed up without more than gentle pressure. We can use the speculum for this operation, or do it without that instrument. The sponge should be mounted on the staff made for the purpose, and passed up so that the upper end reach inside the os internum. Ordinarily it will remain in place without any support, but if it slip out we may place a tampon of cotton secured by a piece of twine beneath it until the sponge begins to expand. The patient may withdraw the cotton in two hours, and wash out the vagina with a weak solution of carbolic acid. In ten or twelve hours the sponge may be replaced by another as large as will conveniently pass in the dilated cervix. This last sponge may remain about as long as the first. If the induration is not considerable, we need not use the large one, as the lesser one will be sufficient. When but one sponge is employed, it ought to remain sixteen or eighteen hours. The patient can remove the sponge by traction upon the twine attached to it. The introduction of the sponge should be done once in from seven to ten days. In six or eight weeks from the commencement of this treatment, it will generally be better to suspend it for four or five weeks, meantime to employ mild stimulants to the mucous membrane of the cervix every five or six days, or the tincture of iron, or a solution

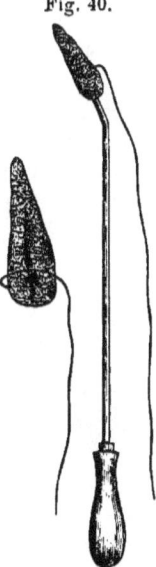

Fig. 40.

of nitrate of silver. If at the end of this time the induration is not decidedly better, the sponges may be repeated. The practitioner should remember, however, that after the absorption has once begun it will generally proceed, and that if the cervix is obviously diminished in bulk and hardness, it will be only necessary to keep the general health good, and use the injections and mild stimulants. Most men of experience learn the great resources of nature in this difficulty as well as others, and take more time and rely more upon the co-operation of the vis medicatrix naturæ. It is not often now that I resort to the potassa or other strong caustics, finding in the dilating sponge a sufficient remedy. When the enlargement and induration are distributed around the whole circle of the cervix, the sponge will scarcely fail to effect the object; and even where these conditions are confined to the anterior or posterior lip of the os uteri, it will generally prove sufficient. We sometimes, however, meet with enormous enlargements of one or the other of these portions of cervix where it is impracticable to introduce the sponge. Other measures, of course, will be required in such instances, and the best is either the white-hot iron or the caustic potassa. The last is the easier used, and if the impression is made with sufficient energy and deep enough, will prove equally efficacious.

Blistering the cervix is another method of resolving the induration of that part. It is a remedy well adapted to the slighter forms of that condition. The blistering may be done with vesicating collodion or hot iron. The latter is undoubtedly the better means, on account of efficacy, convenience, and compara-

Fig. 41.

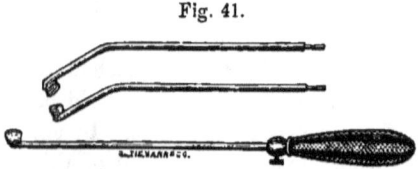

tive painlessness. The irons, of different sizes, made by George Tiemann and represented in the figure, may be heated in the blaze of a gas-burner or spirit-lamp until hot enough to blister.

It need not be red, but only slightly short of that temperature, and applied to the parts to be affected through the speculum. Any speculum large enough to expose the parts well will answer the purpose. Just before introducing the iron, to guard against heating the speculum by contact, we may wrap it quickly with flannel up to the bulb. This precaution will not be necessary if the speculum is made of wood. The iron may be glided over a large part, or the whole of the indurated surface; a blister will be formed immediately if the iron is hot enough, an active inflammation of a moderate grade will occur in the parenchymatous tissue, and in a few days subside, having carried off a portion of the indurating fibrin from the areolar substance of the cervix. In two weeks, and sometimes in one, the vesication may be again produced, and so on repeated until the part is cured. The second and third vesications, and all subsequent ones, will be attended with more effect than the first. If the blistering collodion is used, the parts should be exposed, cleansed, and sponged with sulphuric ether. After the ether has evaporated, the collodion is spread upon the induration, one layer after another, as upon the external surface, until a thick film is formed. This will retain its place until vesication is complete, which will ordinarily take ten or twelve hours. In order to prevent the collodion from coming in contact with the vaginal mucous membrane, we should have several layers of patent lint cut of a size large enough to cover the collodion perfectly, and after having spread the surface with it, a layer of dry lint should be placed upon it evenly, and two or three layers of the same moistened with a weak solution of soda, and all kept in position with a cotton tampon. A drachm or two of sub. nit. of bismuth powder placed up in the vagina against the collodion, will answer the same purpose. Some such expedient should be resorted to in all cases where vesication is resorted to by the use of cantharides, as vaginal inflammation is apt to be a painful accompanying accident. Of course, this process of vesication is superficial in its effects comparatively.

Actual Cautery, or White-hot Iron.—The same irons may be used for causing a deeper impression than blistering, and by destroying the whole mucous membrane and slightly penetrating the submucous tissues, cause a deeper and more extensive inflammation, that will fill all the surrounding parts with effused serum. This latter will dissolve the fibrin that has been deposited in the cellular tissue, and as it is reabsorbed carry the fibrin back into the circulation. For the application of the iron thus highly heated, the parts must be exposed by a large boxwood or horn speculum, and the mucous membrane carefully cleared of the superabundance of secretion, and dried well. The irons may be heated in a charcoal furnace, such as is used by tinners while soldering. They should be so near the operator, that they will not be materially cooled by removing them from the fire. All things being thus made ready, the iron should be taken from the fire and placed in contact with the part to be affected, and held against it with some firmness for five or ten seconds, and then withdrawn. It should not be glided over the surface, but should be held so as to make its effect penetrating and concentrated. If the operation is done with address, no damage will be sustained by any of the surrounding parts, and the patient will experience no pain, but a deep wound will be left, as the effect of the actual destruction of the membrane, over an area something larger than the iron applied. It will require from fifteen to forty days for this wound to cicatrize soundly, and the operation ought not to be repeated until that effect is reached, which may be determined by examination.

The cauterization may be repeated in a proper time thus judged of. The only other thing I need say in reference to the use of the actual cautery, is as to the size of the iron used to produce the effect desired. If the induration is large, the iron should also be large. One-sixth, and still better one-fourth of the surface of the induration on each side should be touched. If the induration is on both the labia uteri, one-fourth or one-sixth of the surface of each should be burned, but at different times, giving the first time to be healed before the other is operated upon. It scarcely ever will be necessary to reapply the actual cautery to the same place. The subsequent ones may be removed from the locality of the first. Generally, one cauterization will answer for one enlarged labium.

Caustic Potash—Object of Using and Mode of Applying.—Too great caution cannot be taken in applying the caustic, for fear

TREATMENT OF SUBMUCOUS INFLAMMATION.   205

of an unnecessary, if not mischievous, extension of its effects to other parts. We should be prepared with the ordinary quadrivalve speculum, a pair of dressing-forceps, some cotton-wool, acidulated water, and sweet oil. The parts ought to be fully exposed, the whole cervix included in the end of the speculum, and illuminated with a good light from a large window, and the patient so placed that we may operate without restraint of any kind. When the cervix is thus included in the speculum, we should take a piece of cotton-wool, as large as may seem to be necessary, thoroughly saturated with acidulated water, and place it beneath the cervix, so as to underlie the part contained within the speculum, and come in contact with the end of the cervix below the point to which we wish to apply the caustic. We should also pour into the speculum a small quantity of the acidulated water, enough to fill up the end of the instrument as far as it can be without being in contact with the part to be operated upon. It is almost, if not quite, impossible to apply the caustic potassa without having it to run more or less. This should be remembered and be provided for. Some liquid, as here recommended, that will immediately decompose this chemical, should be kept in contact with all the parts where there is a possibility of its touching by flowing upon them. I have sometimes, where the direction of the speculum allowed the retention of the fluid, simply poured in the acidulated water until all the parts were inundated that were in danger of injury from the contact of the remedy. After having taken these precautionary measures to secure the surrounding parts from harm, we may secure the caustic in any way most convenient, and apply it as may seem necessary in the case. Ordinarily I seize it with my dressing-forceps, and use them for a caustic-holder. The extent and duration of the application must be determined by the appearances at the time of using it. The enlargement and induration sometimes includes the whole extremity of the cervix, while at others it is wholly or nearly confined to one of the lips of it. I do not think it desirable to apply the caustic extensively. One slough is usually sufficient, and most beneficial, for one application. This should be made in the centre of the indurated part, if the induration be confined to one of the cervical labia; but if the whole extremity is the subject of the induration, the slough may be made in the central portion of the cervical lip, at the upper part of the included portion. The depth of the slough should be sufficient to destroy the mucous membrane and penetrate the submucous tissue. This may be done by holding the caustic steadily in contact with the part sufficiently long. The slough should be not larger than a dime in circumference, and, in cases of moderate enlargement, the size of a half-dime will answer all purposes. With reference to the depth of the impression, I would say that I have oftener regretted having made too light an application, than too prolonged. Thoroughness, combined with carefulness, is just as necessary in the application of the caustic potassa, as of the nitrate of silver. One application is not ordinarily sufficient; but it should not be repeated too near together. The best rule by which to be guided as to the time for repetitions, is to wait until the effect of the first has entirely subsided. This will require from three to six weeks, owing to the extent of slough and the curative capacity of the individual. As soon as the effects of the first application have entirely subsided, we may make a second application, as remote from the locality of the first as the size of the induration will admit. When this has gone through the different phases of inflammation, sloughing, and healing, we may make the third, &c. We may thus repeat the applications until the induration is removed. The time selected for making these severe caustic applications should also have reference to the general condition of the patient, and the special condition of the uterus. It would, of course, be improper to make the application at a time when the patient was in any way predisposed to febrile or inflammatory action, or pregnancy; menstrual congestion should also forbid it. Midway between the menstrual periods, the uterus will most likely be in the best condition. The patient should remain in bed for the first twenty-four hours after this sort of application, and she should abstain from fatiguing use of the lower extremities for at least a week. If there should be much pain, heat, or febrile excitement, which, in my observation, are very rare accompaniments, soothing injections, of flaxseed infusion with laudanum, may be repeated three or four times in twenty-four hours, injections of tinct. opii in the rectum, fomentations to the hypogastrium and perineum, and warm hip-baths. In all cases, where no particular objection exists, the patient should use the sitz baths, and injections of tepid water, twice a day, between the applications. As soon as the slough is completely detached, and suppuration indicates a good condition of the ulcerated surface, the danger sometimes attending these applications is no longer to be apprehended. I have not found it necessary, nor do I think it best, as directed by Dr. Bennett, to dress the place with nitrate of silver applications.

In all these cases of chronic submucous inflammation, we will expedite the cure by maintaining the functions in the most healthy, or nearest to a normal condition, within our power. The bowels should be regulated particularly: they should be free and unloaded. I have never found it necessary to resort to the actual cautery for the cure of these indurated and enlarged cervices. The caustic potassa has been sufficiently powerful for all purposes. Perseverance should be a guiding principle. Twelve months is not a long time to effect a cure when this kind of organic lesion results from inflammation.

CHAPTER XIV.

DISPLACEMENTS, THEIR PHILOSOPHY AND TREATMENT.

From what I have already said, it will be inferred that, for the most part, I consider displacements as complications and effects of inflammation; and that any other treatment but such as will remove the causing condition, is not curative. The idea that they are primary affections, and require independent treatment, is so firmly rooted in the minds of patients and physicians, and the fact that we cannot always cure the uterine inflammation and enlargement upon which they depend, and moreover that they do in very rare instances occur as the effect of apparently inscrutable causes, and consequently call for special treatment, render a separate consideration of these proper and necessary. It is plain that the notions in reference to the main item of treatment, viz., mechanical support, are too vague to be profitable to the novice. It will not be possible for me to enter into a very minute detail of such treatment, nor do I wish to be understood as trying to do anything more than suggest an outline of the principles that should govern us in the use of means for the relief of them. To do this satisfactorily, I must consider succinctly the different sorts of displacements, and, if possible, arrive at their mode of producing inconvenience and suffering, and their mechanical effects upon other organs, especially those in proximity to the uterus. I need not say that such considerations must also be imperfect, and assure the reader that they are merely intended to be suggestive.

Many, if not most of the failures of mechanical support for the relief of displacements, depend upon a want of correct knowledge as to their nature in any given case, and consequently, of the right kind of instrument to be used, and the mode of applying it. It will be found, upon attentive examination of them, that displacements cause distress by pressing or dragging upon other organs, by taking away the support of other organs, and thus allowing them to be misplaced: as when it sinks low into the pelvis, the abdominal organs fall into this cavity to fill the vacuum; or when the cervix or portion of the uterus is tender, causing distress by bearing on these places as it settles against the perineum, sacrum, or rectum. In retroversion, for instance, the round ligaments, broad ligaments, and bladder, are drawn more or less out of their place, and cause suffering; so with all other displacements, varying, however, in their respective effects. In addition to this, the uterus may, in retroversion, press upon the rectum and cause inflammation and pain in it; or, by stopping up this organ, cause costiveness and difficult defecation. Retroversion of the uterus, when the fundus or posterior wall is inflamed, causes suffering by the contact of the tender part against the rectum, sacrum, or perineum.

Nature of Displacements.—The nature or doctrine of displacements of the uterus involves the conditions of its annexed organs. The uterus cannot be retroverted while the round ligaments retain all their natural conditions as to length, position, &c. Some parts of the vagina must be also changed in their conditions. So of prolapse; the round, and particularly the broad ligaments, and the vagina, &c., must be made to deviate by the acting causes of these displacements, before these latter can occur. And it is not a question which is at fault, as though the onus of failure in the functions appertained to either the ligaments or vagina, but which is most at fault. This requires us to decide which one of them does most to hold the uterus in place, which can be done only by distinguishing the kinds of displacements, and considering them with reference to this matter. The vagina can do but little to resist causes operating to produce retroversion or anteversion, but if narrow and rigid, may strongly resist the tendency to prolapse. The ligaments seem so arranged as to resist displacements in every direction, and, excepting anteversion, they are pretty firmly opposed to acting causes. Another point of importance in the doctrine of displacements is, the determination of the question whether the displacements occur as the effect of relaxation of the sustaining organs, or whether the sustaining organs are relaxed on account of the long-continued action of the causes operating on the uterus. In most cases, I think, the last of these conditions obtains; but, undoubtedly, they are quite frequently contemporaneous and consentaneous circumstances. After labor, the ligaments and vagina must have returned to their natural and healthy dimensions and firmness before they can resist displacing influences. This they cannot do for some weeks, and during this time the weight of the uterus, on account of imperfect involution, is much greater than they are in the habit of sustaining. Although usually at fault, because overcome when in its natural condition, and forced into deficiency of function, they may, as the above view shows, be deficient because of their condition at the time the operating causes are applied. But we cannot always say that the uterus has fallen, or become displaced, because the ligaments and vagina are too weak or too lax to sustain it in place, but in many instances because the acting cause is sufficient to force them or overcome them, and carry the uterus in the course of its action in spite of their healthy resistance. I think this is the true explanation in most cases of displacement.

Depression or Lapse.—The principal and only important varieties of displacements are, first, a simple depression or falling of

Fig. 42.

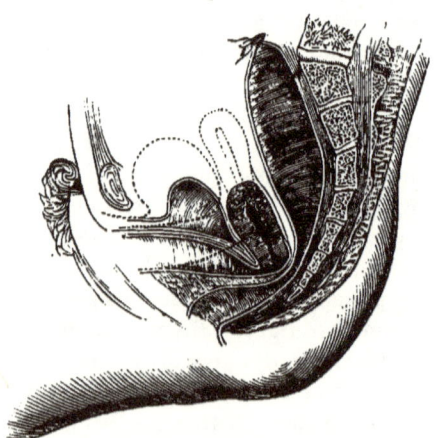

Lapse, or Descent of the Uterus, without change of axis, and the displacement of the bladder and pressure upon the rectum.

the uterus in the axis of the superior strait. The inconveniences resulting from this deviation are painful tenesmus, constipation, or hemorrhoids, on account of pressure upon the rectum; sciatic pains, on account of pressure upon the sacral nerves; pain in the uterus itself, on account of pressing upon its own tender cervix; and feeling of weight on the perineum, and dragging about the loins and hips. The broad ligaments are stretched; perhaps the round ligaments somewhat increased in length, less so, however, than the others. The change in the direction of the vagina from almost directly backward to backward and downward, is also quite obvious. The rectification of this deviation is usually accomplished by lifting the uterus up. It requires no change in axial direction with the pelvis, but simply an elevation to restore the uterus to its proper place.

Prolapse.—Secondly, prolapse in various degrees, from slight depression to complete extrusion from the labia. This displace-

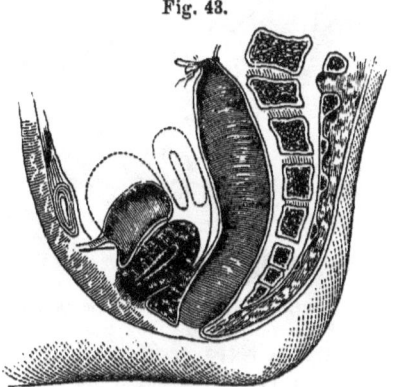

Fig. 43.

Prolapse of the Uterus.

ment in its slightest degree, and in fact in all its degrees, pulls upon and stretches all the ligaments, the broad ligaments by far the most. The vagina suffers displacement proportionally with the prolapse. It is inverted, its walls being doubled upon themselves, and its cavity progressively shortened until entirely effaced. This displacement is always in a direction corresponding with the axis of the vagina, and different portions of the pelvis, and follows the curve formed by the hollow of the sacrum and continued by the perineum. The inconveniences arising from this displacement are not unlike the last, until it becomes great, when the bearing down becomes more distressing, as well as the dragging on the loins. To these are added those arising from a prolapse of the abdominal viscera into the pelvic cavity; sinking sensation in the epigastric region, and dragging upon the hypochondria, &c. The means calculated successfully to restore the uterus in this displacement must lift it up and correct its axial deviation.

Fig. 44.

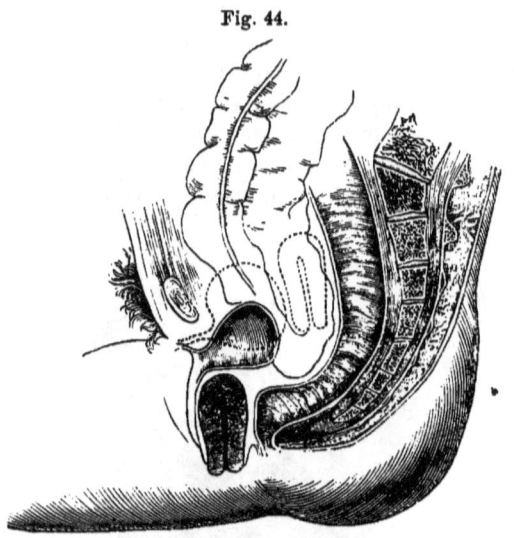

Protrusion of the Uterus, with attendant change in the position of the bladder and intestinal canal.

Retroversion.—Thirdly, retroversion. This displacement is present when the fundus is depressed by being thrown back into the hollow of the sacrum, while the cervix is drawn forward and upward so as to be upon a level, or above a level with the arch of the symphysis pubis. The difference between this and prolapse is, that the fundus is thrown down lower into the hollow of the sacrum, and the axis of the uterus is almost natural, but the lower end becomes the upper. The inconveniences arising from this displacement are caused by pressure on the rectum, perineum, and sacral nerves, in the posterior inferior part of the pelvis, and sometimes pressure upon the neck of the bladder, or urethra in front, and dragging upon the ligaments. The ligaments most severely stretched are the round, the broad being much less so. The condition of the vagina is changed very considerably; the anterior wall is very much shortened, while in married women the posterior is elongated somewhat. The means employed for the correction will act by elevating the fundus and pressing the cervix backward toward the middle of the pelvis.

Fig. 45.

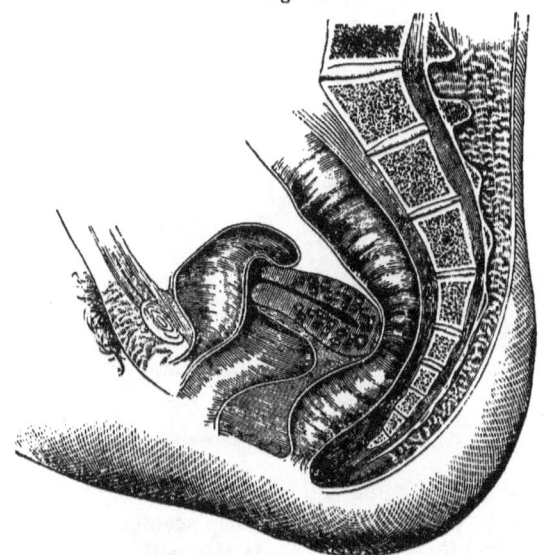

Retroversion of the Uterus, showing the Fundus pressing upon the Rectum, and the Cervix encroaching upon the Bladder.

Anteversion.—Fourthly, anteversion is, in most respects, nearly the opposite in position. The cervix is turned back upon the sacrum, and elevated somewhat above its natural position, while the fundus is thrown forward upon the bladder and anterior walls of the vagina, so as to come down to a level, or nearly so, with the arch of the symphysis pubis. The inconveniences arising from this displacement are caused by pressure upon the bladder, urethra, and rectum, tenderness in sexual intercourse, and dragging about the pubis. The broad ligaments are stretched most, and the vagina is elongated and depressed somewhat at its posterior extremity. Not unfrequently the rectum is pressed upon by the cervix uteri, and distress arises as a consequence. The means for rectifying this position must lift the fundus upward, and push it backward, or draw the cervix forward and lift it slightly upward.

Fig. 46.

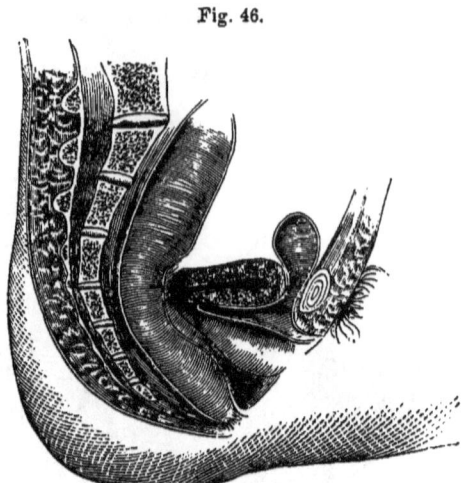

Anteversion of the Uterus.

Causes.—Anything that will increase the weight of the uterus predisposes it to deviations and displacements. When thus predisposed by increased weight, if the patient is much in the erect posture, the uterus will settle down into displacements. It will be observed that the deviations I have mentioned are lapses in some manner or form. When the uterus is slightly enlarged and increased in weight, the erect posture is not always enough to cause displacements; but if the patient strain from the tenesmus of dysentery or dysuria, or in lifting, or is jolted so as to bring the weight of the abdominal viscera down upon the pelvic organs, she feels a distressing sense of pressure upon the perineum, rectum, or bladder, or all of them, and thenceforth she suffers from some of the disagreeable symptoms attendant upon the displacement. Should the uterus be much larger than natural, the erect posture maintained for any length of time insures a displacement. Inflammation of the cervix is almost always attended with increased size and weight of the whole uterus, and thus it predisposes the organ to displacement. This accounts for the fact that we very often find these two conditions present. I have no doubt, from ample observation, that inflammation is very frequently the cause of depression and other displacements in this way. Dr. Bennett thinks that when inflammation attacks the posterior walls of the uterus, retroversion is the result, and that anteversion is caused by inflammation of the anterior wall, and leaves us to infer that these displacements are almost always connected with inflammations as an effect. While I believe with him that these displacements may result from inflammations thus localized, they are, I am satisfied, often caused by inflammation of the cervix alone, and that without any peculiarity discoverable by an examination. I think I have seen every variety of displacement connected with cervical inflammation.

Imperfect Involution.—Imperfect involution is another cause of displacements. Involution may be imperfect, and yet be progressing naturally when a co-operating cause determines displacement. If, for instance, a woman who has given birth to a child, arise from the bed in four or five days, and maintains the erect posture for a length of time, or remain permanently out of bed, engaged perhaps in some arduous duties on her feet, the uterus, being still so much above its natural size and weight, falls in some manner in spite of the relaxed ligaments and vagina. This is the manner in which some persons contract uterine displacements after labor.

Arrest of Involution.—But it occasionally occurs that, long after the usual time for involution to be complete, the uterus remains increased in volume and weight, because of the arrest or the tardiness of this process. In this condition, the uterus is liable to settle below its natural level, in part or as a whole. Subacute inflammation, attacking the post-parturient uterus, not unfrequently delays the return of this organ to its natural dimensions for many weeks after confinement. In all these last cases, the displacement occurs at a period not very remote from parturition or abortion. And very many cases can easily be traced to this time, or near it, seeming to be a sequence to it. I cannot forbear remarking here, with reference even to these cases, that when they have become chronic they are found to be complicated with inflammation; and that where this can be removed entirely, the restoration of the position of the uterus generally takes place spontaneously, and is easily effected by treatment, or ceases to present any indication for treatment, on account of the absence of symptoms.

Tumors.—Tumors, developed in some part of the fibrous structure, cause an increase in weight, and thus encourage, at first, displacement, and after awhile determine it. The position of the tumor will govern the nature of the deviation. If the tumor is in the anterior wall, it is apt to cause anteversion; if in the posterior walls, retroversion; if in the cavity or cervix, prolapse, or merely lapse.

Loaded Intestinal Canal.—With the causes above spoken of predisposing to it, I think the pressure of heavily loaded intestines may determine displacement. Fecal accumulations in habitual constipation would be sufficient.

Distended Bladder.—Equally, perhaps more certainly mischievous, is a bladder constantly filled with urine overriding the uterus, and pressing backward and downward. This cause would seem to favor retroversion, on account of the position of the bladder. These causes are undoubtedly not sufficient, however, to produce permanent displacement, without the co-operation of increased weight of the uterus.

Symptoms.—However the displacements may occur, the symptoms attendant upon them will not enable us to distinguish one from the other. As I have already shown, certain deviations very often give rise to particular symptoms; but this is not so constantly the case that our diagnosis can be materially influenced by them. It is true, also, that the symptoms cannot be distinguished from symptoms arising from other pelvic diseases; hence there is no alternative left us; a physical examination will be our only sufficient means for forming an accurate diagnosis. The symptoms are an expression of the sufferings of other organs, for the most part, from pressure on them by the uterus. Pain, numbness, debility, formication, or change of temperature, general or partial, of the lower extremities, or one of them, on account of the pressure upon some of the large nerves, particularly the sciatic running down them, or tenesmus, constipation, hemorrhoids, and sense of heat or weight in the rectum, indicate the pressure upon that bowel. The dysuria, cutting, burning, or rending pain in the bladder, incontinence of urine, and other distressing vesical disorders, are expressive of the suffering caused by pressure on the neck of the bladder or urethra by the displaced uterus. But a general pelvic tenesmus, or feeling of bearing down, with weight and dragging upon the perineum, not unfrequently are produced or aggravated by the uterus lying heavily upon the bottom of the pelvis. This is perineal distress. There is also a general feeling of pelvic distress, such as dragging pain in the hips and loins, weight and pressure about the pubis, a feeling as of a cord drawing in one or both of the inguinal regions, or general sense of weakness and indescribable *malaise*. Another sort of difficulty seems to be produced by the uterus in its descent dragging other organs out of their natural position. The bladder may be thus drawn down, and cause a sense of dragging from the umbilicus, or produce various sorts of trouble in the functions of holding and evacuating the urine.

Prolapse of Ovaria.—The ovaria are displaced and drawn down into the pelvis, with the feeling of tension in the broad ligaments or iliac regions. In extreme cases of prolapse, and even sometimes in the slighter degrees of displacement, disagreeable symptoms arise from abdominal organs falling into the pelvis to fill the partial vacuum caused by the descent of the uterus. To this cause, doubtless, may sometimes be justly attributed the sense of weakness and emptiness in the epigastrium, and the pain in the side, in the region of the liver and spleen.

What is more common, however, is the suffering caused by the inflamed uterus, slightly depressed, pressing upon its own sensitive diseased parts, as the cervix, posterior wall, or even fundus. Again, there can be no doubt but that the rectum, bladder, cellular tissue within the pelvis, and parts surrounding the nerves, and even the nerves themselves, are often subjects of inflammation, and, being pressed by the inflamed, enlarged, and sensitive uterus, are much more susceptible to the above described influences than if they were all in their ordinary healthy condition.

Examinations to determine displacements should be both

digital and instrumental, and thorough enough to ascertain the particulars as to the position and condition of the uterus in all other respects that can have any bearing upon the case.

When very great suffering of the character above mentioned exists, there is almost always a combination of inflammation and displacement. Should there be any obscurity in the position of the uterus, on any account, our diagnosis may be cleared up to a demonstration by introducing the probe into the cavity of the organ. The direction the instrument takes will clearly show the direction of the uterine cavity, and thus indicate the position of the organ.

Treatment of Displacements.—The efficacious mode of treating displacements, as I have before intimated, is the perfect removal of the causes of them. When this can be done completely, the inconvenience and suffering attributed to them are removed, or very materially ameliorated. But there are cases where this cannot be done from various reasons, among which are the prejudices of patients and medical men against the treatment necessary for the cure of inflammation, the impossibility of curing the inflammation when every opportunity is enjoyed; and even sometimes when the inflammation is removed, so far as we can judge, there may be continuance of displacement with its symptoms. In all, or most of these, the skilful treatment of the cases, as displacements, will often result in great palliation, if not in cure. All I need say here in reference to the treatment of inflammation is, that full directions may be found in the foregoing chapters on that subject.

Treatment of Deficient Involution.—Involution as one of the causes, should be treated as subacute inflammation of the whole of that organ. All the symptoms, as we usually find them, will justify this procedure. It is, in fact, subacute inflammation occurring in the uterus after accouchement, and operates in establishing displacement upon first getting up from childbed. To avoid this condition by sufficient quiet and care is much easier than the cure after the disease is established. When it does exist, rest in the horizontal position, alteratives, laxatives, external fomentation, counter-irritation, &c., will ordinarily relieve it, and prevent or cure the mischief done or apprehended.

Removal of Tumors.—When the uterus is enlarged or depressed by tumors, their removal is the only radical way of overcoming the difficulty. If a loaded condition of the bowels is the operating cause, we must endeavor, by alteratives, laxatives, &c., to remove this condition.

Mechanical Support.—The main object I had in view in introducing this chapter on displacements, was to discuss the subject of mechanical support. And as an introduction, I may state my conviction, that very few general practitioners study these affections sufficiently to acquire the skill requisite for the best management of them. Too often we are satisfied with merely recommending some form of supporter or pessary, and leave the execution of our designs to the patient or her friends. There is too much carelessness in this respect to form a proper estimate of the nature of mechanical support for the uterus. Believing these statements to be true, I have ventured to attempt to describe, in a very concise manner, as far as they have been suggested to me by a large observation and a careful study of the subject, the conditions which give direction to the particular kinds and modes of using these contrivances.

Two Kinds of Mechanical Support.—There are two general kinds of mechanical contrivances for the purpose of support.

Abdominal Supporter.—The first which I shall mention is the abdominal supporter. It is made of various shapes, sizes, and materials. The object is to lift the abdominal organs off the uterus and its appendages, and support them so that they cannot press the latter into the pelvis.

Rationale of their Action.—It will be remembered that the inner face of the pubis looks obliquely upward and backward, forming the base of support for the abdominal organs. These latter press, on account of the inclined position of the pelvis, obliquely upon the uterus, instead of perpendicularly. The farther back the pubis is pressed toward the sacral promontory, the more completely beneath the abdominal organs, the more certainly they rest upon the upper portion of the inner face, and the less they press into the pelvis. Abdominal supporters are intended to, and fulfil their purposes best when they aid the pubis in this function. One indispensable portion of them consists in a plate or pad, so arranged as to press the abdominal muscles immediately above the symphysis as far back toward the promontory of the sacrum as possible. This plate, pushed in above the symphysis and below the intestines, prevents them from weighing heavily upon the pelvic viscera. And when it is recollected that it requires, in most persons, a very little intrusion of this plate or pad into this part of the abdomen to intercept a line falling through the central part of the trunk above, and thus to assume a position for the support of the abdominal organs, we can see how it might be efficacious in certain instances, and nearly harmless in all. The plate or pad is pressed to its place, and held in position by springs or bandages variously arranged to suit the fancy of the contriver. Connected with the springs which pass around the hips, not unlike the springs of a truss, is a pad that bears upon the back, so as to press the loins forward, and thus incline the face of the pubis more to the horizontal. The cases to which properly formed abdominal supporters are applicable are not numerous; but there can be no question about their utility as a means of ameliorating the distress sometimes experienced on account of pressure on the top of the pelvis, and on the pelvic viscera by the abdominal organs. The observation of the profession has not settled into rules as to the manner of using these supporters, nor are they used at all by many of the most intelligent members of the profession. Yet I am persuaded that in certain instances where we are under the necessity of confining our efforts to palliation, the supporter, judiciously selected, adapted, and applied, will afford relief that cannot be obtained by any other means. There are many circumstances and conditions to be studied as indices to the kind of instrument applicable, its mode of use, and as affording data by which to judge of the practicability of any kind. These considerations should be studied in each case. Thin persons with large pelves, who stoop habitually forward, are the sort of patients, as a general thing, whose shape and condition render them applicable; while fat, erect women do not profit by them. These conditions are mentioned irrespective of the special affections for which they are recommended. Their adaptability depends, further, upon the complete absence of tenderness of the hypogastric region, or of great sensitiveness from any cause in any portion of the body upon which any part of the instrument may press. The firmness of the abdominal muscles may also prevent the supporter from having its proper effects. But notwithstanding the fact that the case and the patient may be suitable, so far as we can judge, a trial will be the only means of deciding the adaptation. I hope the profession will try more patiently and philosophically this means of palliating the sufferings of women, when we cannot cure the diseases which give origin to them.

Its Value.—It is needless, after what I have said above, to state that I consider the abdominal supporter but a make-shift, and useful as a palliative only when radical means cannot, for any reason, be used or relied upon. Unfortunately, these are too numerous to be allowed to go unprovided for.

Supporter and Pessary combined.—They are sometimes available as external attachments to pessaries, thus keeping these last in place, and making them more completely fulfil their special purposes. The combined abdominal supporter and pessary, when they can be borne, make an efficient means of rectifying displacements. The perineal pad so often attached to the supporter, although more easily worn and less likely to do damage, is also less efficient than a suitable pessary in the vagina, kept in place by a spring coming down in front from the front pad of the supporter.

Pessary.—The pessary, however, is much more commonly used alone than in combination with any other means of support. This instrument is made of various materials, and in many different shapes; the grand object of all of them is to maintain the uterus in its proper place and position. They are direct supporters of the organ by touching the uterus in some part, and by the contact holding it in place, or support it by converting the vagina into ligaments of support.

Ring Pessary.—The ring pessary, for instance, by distending the vagina all around the cervix, supports the uterus upon a level with its own circle, thus lifting it off the perineum, rectum, or nerves at the bottom of the pelvis. The ring pessary, by pressing upon the vagina, and drawing it upward and backward, is calculated to correct some inclinations of the uterus as well as to lift it up in place, as in cases of retroversion particularly.

Stem Pessary.—The stem pessary, as it is introduced in part into the cavity of the uterus, if properly adapted, also corrects all sorts of deviations. It should have a perineal flat support, or be attached to some external means of maintenance. It is the most perfect mechanical support we can make use of, for if adjusted in accordance with a just knowledge of the natural place and position of the uterus, it is certain to prevent it from departing from it.

Globe Pessary.—The globe lies immediately behind the pubic bone, presses upon the anterior wall of the vagina and uterus, lifting the fundus upward and backward, and, when sufficiently large, raises the whole uterus by drawing upon the anterior wall of the vagina.

Oval.—The oval occupies the same position in the vagina, and operates the same way when it lies crosswise, immediately behind the rami of the pubis and ischium.

Disk.—The disk with convex depression in the centre merely lifts the organ up from the perineum, by its thickness. It lies under the uterus, in the centre of the pelvis, and is perpendicular in its action. It is incapable of correcting any other malposition than depression or prolapse. The gum bag, distended with air, of all these shapes, globular and circular, diskal and oval, of course operates as I have described these. Notwithstanding the above is a general, and, for the most part, correct idea of the modus operandi of the different shaped pessaries, in no two cases will the same formed pessary have precisely the same bearing above and below, and consequently these items cannot be attended to in each different case. There is, in other words, an individuality in every case of every sort of deviation that will require for it separate study. This instrument is not only made in different shapes, but there is a diversity of material used for its construction. Several of the metals, as gold, silver, copper, steel, &c., enter into the composition of pessaries; horn, wood, ivory, India-rubber, and gutta-percha, are also made use of. These two last are worked with so much skill, of late years, that they are taking the place of other material in the manufacturing of pessaries. The hard and soft rubber pessaries are assuming almost every variety of shape. Although it is quite impossible to give directions that will be applicable to all sorts of cases, and very much must always be left to the judgment of the attendant, I venture to hope that the few general considerations which I shall submit will awaken intelligent reflection in the mind of the student as to the difficulties that will often present themselves, and sometimes entirely baffle him in the use of the pessary.

Preparation of Vagina.—It should be remembered that the vagina will frequently need preparation before it will tolerate the pessary. How unreasonable it would be to introduce the globe, oval, disk, or ring pessary into the vagina, and expect it to tolerate the presence of any of them, when it was in a state of inflammation, great contraction, or rigidity. These conditions, if present, should be removed before attempting the use of any pessary. When this is not practicable, we should not think of using the instrument. A condition of the vagina most tolerant of the pessary, but which often thwarts our best considered plans, is a very lax state of its walls or sphincter. This relaxation sometimes obstinately persists, in spite of every effort to remove it. By attaching external supports to the pessary, in these cases, we may keep it in position, and thus compensate for the absence of the co-operating support of the vagina.

Condition of the Uterus that modifies the Use of the Pessary.— These considerations as to the state of the vagina have but little reference to the kind of instrument, either in shape, size, or material; but there is another class of circumstances that will govern us in the selection of a pessary. These have reference to the suffering organ, and the mode in which the suffering is produced. Where is the pressure? What organ is pressed upon, and by what part of the uterus is the pressure made? Does the rectum suffer by the pressure of the cervix, as the uterus stands in the direction of the axis of the superior strait? If so, the uterus must be lifted clear of it by an instrument that will not press upon the rectum in the same place, or the symptoms will be continued, and even aggravated by the pessary. If the rectum suffers by the fundus turning backward upon it, the fundus must be raised forward, without making the rectum a point of support for the instrument which does it. In like manner, if the pressure is upon the sacral nerves, perineum, bladder, &c. In any or all of these cases the uterus may be tender or inflamed. If this is the case, the pessary must be so constructed as not only to avoid pressure upon tender points of the subjacent organs, but likewise to impinge upon the uterus at a sound portion, or support without touching it at all, else the symptoms will be only partially relieved, or changed somewhat in their character. When it is remembered that all of these conditions are to be fulfilled in order to get a perfect adaptation of the pessary, it will not be surprising that we so often fail in getting good from its use, that there is so great a variety in shape, consistence, size, &c., that definite rules are impossible, and that so few practitioners agree in regard to their usefulness and adaptation. The pessary must be studied as a mechanical instrument, while its use subserves physiological purposes. It must be governed by mechanical laws, with the infinite and inappreciable exceptions which physiology always imposes upon them.

Kind of Displacements to which it is Adapted.—Another set of considerations must have reference to the mere displacement; as to whether it is retroversion, anteversion, prolapse, lapse, &c. On these last considerations will, more than any others, depend the shape of the pessary. It will sometimes be found difficult, if not impossible, to employ an instrument that will correct displacements without its making pressure on the suffering organ at the side of, or beneath the uterus, or upon a tender point in this organ itself. When such is the case, the consistence or hardness is a matter of much importance. A pessary filled with air or stuffed with hair is a better point of support for a tender uterus than a hard rubber, glass, or metal instrument. The former kind are cushions of such softness as sometimes to be tolerated by a very tender organ. A deliberate attention to the above considerations will enable us to approximate more nearly to an adaptation than a more loose and less methodical study of each individual case; and while it may not lead us, at the first experiment, to perfection in this respect, it will form a basis for intelligent experimentation. And it should be expected, that not only will such study, but observation also in each instance, be necessary to arrive at a perfect adaptation.

The instrument which, mechanically speaking, is best calculated to correct all sorts of deviations, as well as to keep clear of surrounding organs, and consequently not to cause distress in them, by pressure upon them, is the stem pessary with external support. It in fact completely fulfils every mechanical indication in any species of deviation. This is the case, because the stem fitting into the cervical cavity, and even passing up into the cavity of the body, forms a lever when fixed in a certain position by a branch passing out between the labia to be connected with a fixed support externally, and it must keep the uterus precisely in place. Unfortunately for our success in these cases, this mechanical contrivance is quite intolerable under most circumstances. The reasons why it is intolerable are that the pressure of the stem upon the mucous lining of the cavities causes inflammation, and a positively fixed state of the organ is unnatural, and in some postures and movements of the body must be annoying to other organs by interfering with their mobility.

The elastic ring made of a watch-spring covered by guttapercha, or some other impervious material, when properly adapted in size and strength to the size of the vagina, and well applied, is also a very efficient instrument, and applicable to almost all varieties of deviation. It spreads the vagina out on all sides, and causes the walls of this tube to assume almost the same relation to the lower part of the uterus that the broad ligaments do to the upper part of the organ. This imitation of the pelvic circle by the ring and its ligaments, stretching the vagina around the uterus, keeps the cervix in position, provided the vaginal walls are made tense by the size of the ring. When the broad and round ligaments are lax, however, the fundus and body are left to topple over backward to the sides or in front, and if these are heavy it may distress the rectum or bladder. The ring pessary may be made to replace the uterus in retroversion or prolapse, better perhaps than any others, and is not calculated to be of any advantage in anteversion. It is not so likely to produce intolerable irritation as the stem pessary, and in fact may be made to agree with cases as well as almost any other kind of instrument. The size of the ring to be used will depend upon the size of the pelvis and the tension of the vagina. If the vagina is cylindrical, firm, and elastic, not rigid, free from inflammation and not particularly sensitive, we may hope to procure toleration for the ring. It must be placed so as to suit the case. If the deviation is prolapse, the ring should be placed so that one part of its circumference be at the arch of the symphysis, while the other side is directed to the sacrum, so as to correspond with the axis of the pelvis upon a level with the lower part of the symphysis. For retroversion the posterior part of the circle should be directed up the sacrum toward the promontory, while the anterior part is placed below the arch of the symphysis. It will be seen that the indications are fulfilled by thus accommodating the position of the uterus by the position of the instrument. In the latter position the fundus of the uterus is raised up by the ring pressing the posterior cul-de-sac of the vagina up behind it, while the os is drawn backward by traction on the posterior wall of the vagina. In retroversion the stretching of the vagina from before backward is usually sufficient, and it is not necessary to distend it laterally to the same extent if at all.

Dr. Hodge's modification of the ring pessary, called by him the open lever, may often be substituted for the ring, and in some cases, perhaps, acts better than it. Dr. Hodge's pessary is substantially a flattened ring. It is made of firm material, and curved so that it may be made to distend the posterior vaginal cul-de-sac by curving up behind the uterus, thus lifting the fundus and drawing back the os uteri. This pessary, according to the inventor, is capable of doing more good, and is of greater extent of adaptation than the elastic ring. As it is in the hands of an intelligent and discriminating profession, these assurances will be tested and decided upon, no doubt correctly. I confess that I am decidedly in favor of the elastic ring, which, if not too rigid, moulds and adapts itself to the inequalities of the parts, allows a limited movement to the uterus, and yields to the passage of fœces down the rectum; none of which things are accomplished by Dr. Hodge's pessary, which is unyielding and fixed in its position, and inelastic in composition. Dewees's modification of the ring—the disk—is more clumsy, and, I think, less useful than the ordinary ring of some elastic material. I have seen this form of instrument made of hollow elastic material, and supplied with a tube for inflation. The instrument thus formed is introduced, placed in position and then inflated. Or it is sometimes inflated when manufactured, and kept so permanently, and used as the hard rubber or glass made by Prof. Dewees. The common air-bag, of different forms and dimensions, is made with a supply-tube for filling. This air-bag forms a soft cushion, upon which a tender uterus may rest without much offence to its susceptibilities. It also diffuses the pressure over a large space, in such a way as to relieve the distress of other inflamed and injured organs. There are other forms of pessaries recommended for, and perhaps have, their special virtues; but it would be both profitless and tedious to enumerate more of them. The globe pessary is best adapted to the correction of anteversion, and perhaps none other will answer so well. It is not appropriate in any other form of displacement.

It is necessary to understand the principles which govern their application: study well each case, and then, if there is no instrument at hand that suits, make such a one as is adapted to the case for which it is needed. There are but few persons, who will take time to think upon the instances in hand, but will be able to judge of and adapt the proper instrument. It is a question, after we have adapted an instrument in a favorable manner to a case suited, what should be the management of it. The kind of instrument, and the nature of the case, must determine this question instead of an arbitrary rule. A pessary made of porous material, that entangles the secretions, will soon become foul with them, and hence should be often removed and cleansed. This should be done, for instance, every twenty-four hours. In case of the ring made of hard and polished material, there is no need of frequent removals, as they do not absorb or entangle the mucus, pus, or blood. A profuse discharge of blood, mucus, or pus, will render an instrument that is capable of becoming so, fit to remain in for only a short time. Much tenderness and inflammation is a good reason for keeping both the pessary and vagina clean; but the same states forbid the frequent removal and introduction of the instrument, on account of the violence thus done. The presence of the pessary need not prohibit the use of medicated injections in the vagina; or, if we think best, the pessary may be medicated, or made the medium of applying medicines to the vaginal membrane. Of course the composition of the pessary should be taken into consideration, lest chemical reaction between it and the injected substance occurs. It should not be incompatible with the nitrate of silver, acet. lead, sul. zinc, or with whatever may be used. Carelessness or ignorance, or both, in the use of the pessary, may lead to disastrous damage from them. They should be properly attended to. When allowed to remain too long in the vagina, charged with blood, mucus, or pus, which becomes entangled in them or detained about them, its decomposition becomes a source of poisonous filth, that, by absorption, may endanger the life of the patient, and most certainly will be an intolerable annoyance. Or, by continued and prolonged pressure on some particular place, the pessary may cause ulceration to destruction of much tissue, creating fistula, urinary or fecal. It is but proper and just, however, to observe in this connection that these results are the effects of the abuse of the pessary, and should not be brought forward as objections to its judicious use, any more than cutting the intestine should be regarded as prohibitory of operations for strangulated hernia. Carefulness will, no doubt, prevent all the evil effects which have been done by the pessary; and when the damage cannot be avoided, no judicious practitioner will persist in their use. They are not, in such cases, the means adapted to the end.

APPENDIX.

CASE I.

Case of Ulceration of Os and Cervix, with Inflammation extending into the Cavity of the Neck, attended with Chronic Diarrhœa.

Mrs. B., aged thirty; mother of three children and subject of several miscarriages; has been delicate from childhood. Her bowels have been very susceptible to causes of irritation. For two years before I saw her, she had been subject to almost constant diarrhœa. One remarkable feature of this diarrhœa was that it was attended with a great deal of tenesmus, and not unfrequently with the thin watery discharges which constituted most of them; there were muco-fibrinous moulds cast off from the rectum several inches long. She suffered great pain in the back and left iliac region, and sense of "bearing down" when she attempted to walk. The distress was so great upon attempting it, that she could bear to walk but a very short distance. Her sufferings were increased greatly at the approach of each menstrual period, and were relieved only after the flow was well established. When I first saw her, October 12th, 1858, she was unable to sit up but a small part of the day. She had very slight discharge of pus-colored mucus just before and after her menses; no considerable leucorrhœa at any time. She had aborted about ten months before, and since that time her symptoms had been about as they were then. A digital examination showed the uterus to be somewhat enlarged, and so low as to rest upon the lower part of the sacrum, and pressing heavily upon the rectum, still retaining its ordinary relation to the axis of the superior strait; it was not prolapsed, but simply depressed. When exposed by the speculum, the cervix appeared congested and one-third larger in circumference than natural. At the end, around the mouth, and extending over a large portion of the everted labia, was a scarlet-red patch with small granulations, which bled when pressed by the end of the probe, covering all this, which had to be removed in order to get a plain view of the surface; and extending into the os uteri was a thick, very tenacious mucus, stained yellow with pus. This was so tenacious as to require some time to remove it with the dressing-forceps and cotton. The probe experienced such resistance at the upper part of the cervical cavity as to assure me that the inflammation did not extend through the os internum into the cavity of the uterus. This was a case of simple mucous inflammation with destruction of the epithelium, ulceration of the os and cavity of the cervix uteri. The treatment consisted of the tepid hip-bath, alum-water injections twice a day, and the application of the solid nitrate of silver once every six days, except during her menstrual periods, for about four months. The only other treatment resorted to was anodyne and astringents, to control the bowels when the diarrhœa was too copious, and care as to diet on account of indigestion.

This patient recovered her health entirely, so that she could walk long distances without pain or other inconveniences indicating uterine trouble; she could digest well, and her diarrhœa left her.

For six or seven weeks after the commencement of the treatment my patient improved but very little, except in the gradual subsidence of the pains, and it was not until the treatment by the nitrate was suspended that she improved rapidly. From this time she grew strong and healthy in a short time, and now she is in the enjoyment of better health than ever before. She has passed safely through one pregnancy and confinement, and is now in the sixth month of the second. In each case there were several symptoms indicating danger of abortion, which required some opiates and several days' quietude to allay. The distressing diarrhœa in this case seemed to me to depend upon indigestion and pressure upon the rectum by the depressed uterus.

CASE II.

Ulceration — Severe Constitutional Suffering — Sterility — Complete Cure with Local Treatment alone.

Mrs. P., married; aged thirty-two years; the mother of two children, the last of whom was born eight years before; had from that time symptoms of inflammation of the cervix uteri. For the last three years before I saw her, she had been, as she expressed herself, a worthless sufferer.

She had profuse leucorrhœa, of yellow color and thick consistence; menorrhagia; constant pain in the back and limbs, and a distressing sensation about the perineum. Unable to walk without increasing her pains and other distressing symptoms, her general health had suffered severely from inactivity. She was emaciated, dyspeptic, exceedingly nervous and melancholic. She had had some treatment with nitrate silver, and been subjected to the regimen and all the management of a water-cure; pronounced cured, and returned home. A few weeks after her return to her family dispelled her dream of health, and precipitated her into her former state of valetudinarianism, with a more desponding state of mind than ever before. At this time, April 10th, 1859, I was called to see her, and after gaining the above information in reference to her condition, requested a physical examination of the uterus. This organ was enlarged to double its natural size, the cervix directed backward, was low down and rested upon the rectum. The cervix was somewhat tender to the touch, and the os patulous enough to admit the finger nearly an inch. When exposed by the speculum, the inner surface of both the labia of the os uteri were found in a state of granulation, scarlet color, and the epithelium was so delicate that the slightest touch with the probe was followed by hemorrhage. When thus extended by the speculum, and wiped with cotton so as to be divested of its covering of muco-pus, this patch of ulceration seemed as large as a half-dollar silver piece, and the boundary between it and the healthy mucous membrane definitely marked.

This case was treated by applications of lunar caustic, repeated every six days, made to all parts of the inflamed surface inside the os and out; sitz baths and astringent injections morning and evening. No general treatment of any kind was resorted to from beginning to end, yet the patient was restored to perfect health in about eight months from the time I first saw and commenced directing her treatment.

The sterility, which had been eight years in duration, gave place to fertility, and she has since borne two healthy children. The change in this respect was remarkable, from the fact that conception had not taken place for the eight years preceding her cure, and that in just ten months from the time she was discharged and entered upon a tour with her husband, she gave birth to a son. Conception must have succeeded her complete restoration in about a month.

CASE III.

Endocervicitis without visible Ulceration—Severe Symptoms cured by Local Treatment alone.

M. M., aged twenty-eight years; unmarried; nervo-sanguine temperament; long an invalid (ten or twelve years) from dysmenorrhœa. I was called to see her while she was on a visit to the city from the country, September 18th, 1859, on account of the severity of her sufferings at the time of menstruating. The family with whom she was visiting were alarmed, and believed her to be almost ready to die when they sent for me. I found her suffering most excruciating pains in the hypogastric region, radiating to the back, down the thighs to the hips, and up the abdomen, &c., returning at intervals of from three to five minutes, and leaving her only partially relieved in the intervals. She had cold extremities, nausea, quick pulse, profuse perspiration, and other evidence of severe congestion and nervous suffering. It required large doses of opium, external heat, friction, and volatile stimulants to subdue such great suffering. She said this was a fair specimen of her suffering during each menstrual visitation. She had, of course, had all manner of treatment but local. The main item recently, however, was large potations of gin, to allay the pain at the time of the painful attacks. She was emaciated, dyspeptic, sallow, melancholy, and despondent, and had about given up the idea of ever being well.

In the intervals between her menses she suffered considerable pain in the back, left iliac region, left mammary region, back of the head, and a burning or sense of heat on the top of the head. I ventured to express the opinion that she had disease of the uterus, and that it most likely could not be cured but by local treatment, and explained to her friends what the kind of treatment would be.

After she returned to town for the purpose, I instituted an examination in order to ascertain the precise condition of the uterus. I should state that she had not then, nor never had had leucorrhœa. The menses were correct in periodicity, quantity, and duration.

The finger introduced into the vagina found the cervix low down, and very far back, resting against the sacrum, and turned strongly to the left side. The uterus was large, but not tender to the touch. With two fingers of the right hand under the cervix in the vagina, and those of the left hand pressing down toward the pelvic cavity above the symphysis, the uterus could be included between them, and pushed upward and downward alternately. Upon more accurate examination, the fundus of the uterus could be very plainly felt in the hypogastric region. The sides of the organ were surveyed some distance up the pelvis, but no tenderness was complained of, nor could any protuberance be found. Through the speculum a small amount of thick muco-pus was found to occupy the upper part of the vagina about the cervix, and could be seen in and issuing from the os uteri. The mucous membrane of the vagina and the portion covering the cervix was of a dark livid color. The neck was long, pointed, and rather small; the os rather less than ordinary, and free from abrasions or any other morbid appearance save the color. And this I did not believe indicated inflammation of the mucous membrane, whence pus could be derived.

When the probe was introduced high up into the cervical

cavity, through the os uteri, the patient complained of great tenderness, smarting, and a sense of burning. This burning and smarting continued some time after the probe was withdrawn. This touch of the probe increased the backache and pain in the iliac region. The patient at each introduction seemed very much to dread the increased suffering caused by the probe. The withdrawal of the instrument was followed immediately by the appearance of blood in the os uteri. The "sore place" was about an inch and a half inside the os uteri. The uterus was so large, and the cervix directed so strongly backward, and was so low down, that I had considerable difficulty in exposing the parts perfectly. The os uteri was smaller than usual, and the cavity of the cervix also less than common. My opinion, after this examination, was that there was inflammation of the mucous membrane of the cavity of the cervix uteri; that it was probably the cause of most, if not all of the local symptoms, and the general condition of the system, and that the cure of it was indispensable to the recovery of health. The lunar caustic was introduced far up as possible into the cervix by means of the flexible caustic-holder, once a week, unless when, as it was four or five times, alternated with the acid nitrate of mercury. This last was carried up by the little swab of cotton on a fine piece of whalebone. Both were invariably introduced far up as they would go; above the tender place. Before the close of the treatment, the os uteri contracted so that it was necessary to use the slippery-elm bougie to dilate it, before the swab or caustic would pass up. The nitrate of silver was thus applied three times from the beginning of the treatment before the succeeding menstrual discharge appeared, and the relief from her habitual, excruciating suffering was so nearly complete as to astonish the patient and friends.

From this time forward the distress in menstruating was becoming less, until, at the end of the treatment, this function was in every respect normally performed. Her general health improved uninterruptedly, so that in a year after commencing treatment, instead of being dyspeptic, emaciated, melancholic, she was in cheerful health. This patient, notwithstanding the great and favorable change wrought in her condition, had nothing but local treatment; no kind of internal remedies whatever were used. I should have stated, that she took injections of cold or tepid water per vaginam, and hip baths every morning and evening.

CASE IV.

Case of Ulceration of the Os and Cervix Uteri, Enlargement, and Retroversion, with some Tenderness, without any Local Symptoms, and with an exaggerated state of Nervous Excitement.

Mrs. McD.; aged twenty-nine years; mother of three children, the last of which was fourteen months old. So far as she could remember, her menses had been perfectly regular, as to quantity, quality, and time, except during her pregnancy and nursing. She had never suffered from pain during this process; never had pain in her back, limbs, or iliac region. She had no leucorrhœa, nor, so far as I could learn, any local symptoms indicating uterine inflammation. She had been subject to nervous headache since early girlhood, occurring at irregular times. Her bowels regular, digestion good, had good appetite, and ordinarily slept well. She was rather emaciated, nervous, feeble, and incapable of much exertion; was easily fatigued, and subject, as the consequence, to nervous headache. Her greatest suffering was intense, excruciating headache, in the back and upper part, beginning the day before the appearance of the menses, and growing worse for two or three days, until convulsions supervened. These last would continue, with intervals, for several hours, sometimes twenty-four, and even thirty-six, when the convulsions and pain would gradually subside together. This always produced prostration, that lasted for several days more, so that she hardly recovered from the effect of one attack before another came about.

Upon a careful inquiry into all the symptoms, I could find no apparent cause for so much nervous suffering, but suggested that they might possibly have their origin in disease of the uterus, at the same time stating that I had no other reason for believing so, than their occurrence, in such overwhelming intensity, at the time of the menstrual congestion.

An examination of this organ was instituted, which showed it to be retroverted, somewhat enlarged, and slightly tender to the touch; the mouth patulous and rough to the feel.

There was a large patch of ulceration covering the inside of the posterior lip. A small amount of muco-pus was also found in the vagina, about the cervix and on the surface of the denuded part. The probe did not pass easily into the cavity of the corpus uteri. The treatment in this case was conducted, as usual, by directing baths and injections daily, and nitrate of silver applications once a week. The first effect of it was to cause the ordinary local symptoms of inflammation, leucorrhœa, pain in the sacrum, diminished ability to walk, which before had been good, so far as local distress was concerned. These soon began to subside, however, and with them the excessive nervous irritability. Her menstrual week became a comfortable one, as it had been; her embonpoint was restored. In fact, her general health improved until she was well again, being subject only to her occasional attacks of nervous headache, occurring irregularly, as they had before the terrible menstrual cephalalgia and convulsions made their appearance. This was a remarkable case, on account of the entire absence of the local symptoms ordinarily connected with and dependent upon cervical inflammation and ulceration, the very grave nervous disorders, and their perfect subsidence as soon as the local lesions were removed. No attention was paid to the position of the uterus, and it remained the same the last examination as it was at first. Whether it was rectified afterward, by a gradual return to its natural place, I am unable to say. In other cases I have noticed, in which there was almost, if not entire absence of local symptoms, but present severe general distress, there was the same slight amount of mucus and pus. On the other hand, there is apt to be more of these in cases where local suffering is considerable.

CASE V.

Ulceration of the Os and Cervix Uteri, with Inflammation extending through the Cavity of the Cervix into the Cavity of the Corpus Uteri.

Mrs. K., aged thirty-three years, a strong, plethoric woman; dark complexion; nervo-sanguine temperament; mother of five children; had aborted, about the end of the third month, in four successive pregnancies. In each case the flooding was excessive, and the process protracted, on account of the rigidity of the os uteri. for several days, with a good deal of suffering.

I attended her in the two last of these four abortions. After the third miscarriage, she desired me to endeavor to prevent, if it was possible, the fourth. As soon as she was aware of the existence of pregnancy, she sought my advice. In the preceding two instances she had tried quiet, low diet, and opium for pain, when it first made its appearance. I advised her, in this case, to be more quiet, to keep her bowels, which were habitually constipated, soluble by taking Congress water every morning; if pains indicating uterine contraction at any time occurred, to take opium enough to relieve them; and about two weeks before the time when she might expect an effort at abortion, to be bled, to about ℥xx, from the arm. This treatment was all conducted under direct supervision, but at the usual time miscarriage took place. After recovering from the loss of blood, and the other more immediate effects of this abortion, she was not as well as before it. She complained of backache under exertion, some sense of weight in the pelvis, constipation, and indigestion. She had no leucorrhœa—no irregularity or abnormity of the menses. So little inconvenience was experienced, however, that she thought herself well, except the obstinate tendency to abort, and a little nervousness. As I expressed a belief that she might have inflammation of the cervix uteri, it was thought best to have a speculum examination. The uterus was low down, and the cervix directed backward. By placing the fingers of the right hand under it, and lifting it up, the fundus could be felt, by the left, above the symphysis, and, by the best calculation I could make, the uterus must have been double its ordinary size. There was no tenderness to the touch. When exposed with the speculum, the whole lower end of the cervix was found granulated, scarlet-red, and covered with a thick, glairy mucus, stained with pus. This thick and very tenacious mucopus filled up the mouth of the uterus. Upon introducing the probe, it passed upward about three and a half inches without meeting with any obstruction, or apparent constriction, in any of its course. I believed that in this case the inflammation extended through the cavity of the cervix to the cavity of the body of the uterus, and that probably this condition of the mucous membrane rendered it unfit for the changes necessary for the accommodation and development of the ovum, or evolution of the decidua, in all its perfection. Or that it might exaggerate the natural congestion of pregnancy into diseased excitement.

This case was cured by sixteen applications of the nitrate of silver, with my flexible caustic-holder. The nitrate was applied to all the visible inflammation, and then introduced through the cavity of the cervix into the cavity of the corpus uteri, to the fundus, every six days. After eight or ten applications, the internal os uteri contracted so as to prevent the caustic from passing into the cavity of the body. In all the applications it was introduced as far as it would go, and allowed to remain in contact with the parts several seconds, as it passed along. Soon after a cessation of the treatment she became pregnant, and passed through it without any inconvenience whatever. Her accouchement and getting up were accomplished also in the happiest manner. To the inflammation in this case, I think may be fairly attributed the habit of abortion, as it subsided at once when this was removed by treatment. The complete immunity from threatening or suffering symptoms was remarkable, after so many abortions.

APPENDIX. 239

CASE VI.

Severe External and Internal Inflammation of the Cervix, that would not bear Nitrate of Silver, cured by the Substitutes for it.

Mrs. F., aged twenty-eight, came to Chicago from a neighboring state to consult me in reference to her health, April 2d, 1862. She was very weak, nervous, and miserable, to use her own expression, and for days together was under the necessity of keeping her bed. To come to something of a definite description of her condition: She was very much constipated, and suffered intense pain in the rectum when she had a passage from her bowels. Her digestion was so bad that she could bear but very few of the plainest articles of diet, and often vomited them soon after eating. Every few days she was visited with excessive headaches, mostly in the coronal and occipital regions, attended with prostrating nausea and vomiting. She had constant pain in the back, over the loins and sacrum, pain in both iliac regions, sense of bearing down, or perineal tenesmus; pains in the thighs and hips, and almost constant painfulness in urinating. She was so weak that she was unable to walk a single block, and walking even the shortest distance gave her much suffering.

Her menses returned every three weeks, and lasted ten days. They were very profuse, and she kept her bed always during the time of their flow, and for several days after. In the interim of the menses a profuse yellow, and sometimes greenish leucorrhœa compelled the use of several napkins in the twenty-four hours. A part of the leucorrhœal discharge was very thick and tenacious, and the rest was thin, or about the consistence of pus. Her disease had been of long standing, and she thinks she was the subject of painful and unnaturally profuse menstruation before her marriage, which occurred eight years before I saw her; but her suffering had been very much increased from the time of the birth of her only child, a daughter, now three years old. For the last three years, she thought, she had been pretty constantly as ill as when she visited Chicago. She was very much emaciated; had a cough, and her friends thought she was tuberculous. Her chest was examined, with a view of determining the truth of that apprehension, but no sign of disease of the lungs could be discovered by the stethoscope. There was a peculiar tenderness of the whole abdominal region, manifested more by slight than deep pressure.

An examination per vaginam, with the finger and speculum, gave her great pain, on account of the tenderness near the external orifice. The whole vaginal cavity was intensely red and tender, as was also the mucous membrane of the cervix. The cervix uteri was very much enlarged; must have been, I think, one and a half inches in diameter, in every direction. It was so covered up and besmeared with a thick, tenacious, pus-colored mucus, nothing could be known of its condition until this was removed. After a time, and the use of the dressing-forceps and plenty of cotton, it was laid bare. The whole of its lower end was studded with innumerable granulations, many of them large as a duck-shot, denuded of its epithelium, and bled profusely upon slight pressure. The os was very patulous, allowing the index-finger to pass to the end of the first phalanx, and stuffed with a pus-colored mucus, so tenacious as for ten minutes to defy well-directed efforts to remove it. When it was removed, together with the blood and mucus in the upper part of the vagina, the ulcerated surface could be seen to extend into the cavity of the cervix, as far as it could be observed. When the probe was introduced into the uterus, it passed up three inches without the slightest resistance, and its withdrawal was followed by a stream of blood. The uterus was low down in the pelvis, in a state of lapse, the cervix pressing upon the rectum, and imbedding itself into that organ; while the uterine axis corresponded to the axis of the superior strait, or nearly so. The whole of the lower end of the cervix, and inside of the os, was freely touched with the solid nitrate of silver. She was directed to use injections of one drachm of powdered alum to a quart of water, twice a day, and take a tepid hip-bath as often. Upon visiting her the second day afterward, I found her having profuse hemorrhage, and a great aggravation of pain all over the lower part of her person. Notwithstanding the use of ice externally, pieces of ice passed into the vagina, perfect quietude, elevated position of the hips, &c., for one week the hemorrhage, pain, and soreness had not subsided sufficiently to have another examination. On the twenty-second day after the first application, all the parts in view by the speculum were freely covered with tannin, in powder, and the same substance was carried up inside the os, with the small swab of cotton on whalebone, as described in the proper place. On the 25th of April, caustic potash was applied to the cervix by means of the cotton swab, so as to strongly stimulate the ulcerated surface. It was also carried up into the cavity of the cervix. The parts were dressed with powdered tannin every fourth day until the 10th of May, when the nitrate was again applied. This increased the pain, and caused so much flooding that I did not again venture to use it in the case. Tannin, every fourth day, was used until May 25th; the parts were freely touched with acid nitrate mercury. This did not cause any pain, and was not followed by hemorrhage. From this time forward, the tannin was used every fourth day, except once in about four weeks; when the parts were freely touched with acid nitrate of mercury, and only once more the caustic potassa. The improvement was quite soon perceptible, and in the latter part of November, 1862, she was dismissed, entirely cured of the leucorrhœal discharge, dysmenorrhœa, ulceration, enlargement of the uterus and tenderness; and at present writing, July 2d, has entirely recovered her health, being better than for ten years previous. This patient could not and did not use injections, sitz baths, or medicines of any kind; but was cured by the treatment with tannin, acid nitrate of mercury, and caustic potash, employed with the speculum. It cannot be said that anything else did any good. It was remarkable that she was so intolerant of nitrate of silver as to be made very much worse each time it was used, and yet the caustic potash and acid nitrate of mercury were not followed by inconvenience of any kind. The intense vaginal inflammation subsided also, under this treatment, without anything special for it.

INDEX.

Abdominal supporters, pessaries, etc., 62
 supporters in disease of cervix, 219
 rationale of their action, 219
Abortion, bad management after, 64
Abortions, 59, 64
 conditions of the uterus in, 59
Acid nitrate of mercury and of silver, 183
Accident in injection, 163
Accompaniments, sympathetic, of uterine disease, 26
Accompanying manifestations of moral and intellectual perverseness, 34
Action of instruments represented in Sims's method of examining the uterus, 119
Actual cautery, 203
Acute cellulitis, 80
 inflammation, 129
 supervention of, 129
Affection of the liver, sympathetic, 28
 nervous system, sympathetic, 29
Affections of the spinal cord, 32
 of the sciatic and anterior crural nerves, 33
Aged, appearance of the cervix in, 119
 os uteri in, 107
 endocervicitis in, 95
Amenorrhœa sometimes results as an effect of inflammation in the cervix uteri, 56
Amount of leucorrhœa not always proportioned to extent of disease, 51
Anæmia, 139
Anæsthesia, 34
Anodyne, astringent, and alterative ointments, introduction of, 165
Anteflexion of uterus, 88
Anteversion, 86, 213
Aphthous inflammation, 100, 183
Appearance of secretion, 120
 of the os and cervix, 118
 in the virgin, 118
APPENDIX.
 Case I, 229
 Case II, 231
 Case III, 232
 Case IV, 235
 Case V, 237
 Case VI, 239
Atrophy with hardness, 92
 as the result of inflammation, 124

Bad management after abortion, 64
Baths, 154
 hip, or sitz, 156
 shower, 157
 sponge, 157
 temperature of, 156
Bearing down not caused by displacements, 52
 pain, weight, or uterine tenesmus, 49
Bladder and rectum, extension of inflammation to, 88
 distended, 216
 pressure upon, 87
Blistering the cervix, 202
Bodies, mammary, 40
Bougies, 189
Bowels, sympathetic disease of, 27
Bromide of potassium, use of in nervous excitement, 139
Buttle's uterine scarificator and leech, 196

Can the inflammation be always removed, 71
Cases complicated with phthisis, 76
Cause of cellulitis, 83
Causes of rectal diseases, 84
 of displacements, 214
 of dysmenorrhœa, 55
Caustic applications, diagnostic effect of, 122
 potash, object in using, 204
 mode of applying, 204
Caustic-holder, flexible, 172
 introduced into cavity of uterus, 175
Cauterizing irons, 202
Cavity of the cervix, 93

INDEX.

Cavity of body of the uterus, 93
Cellular tissue, suppuration in, 81
Cellulitis, acute, 80
  cause of, 83
  chronic, 81
  diagnosis of, 81
  extent of, 81
Cephalalgia, 30
Cervical and uterine cavities, length of the, 111
Cervicitis, menorrhagia frequent in, 56
Cervix, blistering the, 202
  cavity of the, 93
Change of general circumstances only temporary in their effect, 128
Character of the mucus in the vagina, 50
Characteristic signs of inflammation, 121
Children, vaginitis of, 66
Childbearing woman, uterus of, 107
  women, external inflammation, combined with internal, in, 95
Chronic cellulitis, 81
Circulatory system, 36
Cold, 62
Comparison of symptoms of uterine disease and spermatorrhœa, 22
Complications, 124
  with cellulitis, 80
    cystitis, 124
    displacements, 85
  of inflammation of cervix, 78
  of mucous and submucous inflammation, 99, 123
  with prolapse of rectum, 84
    fistula in ano, 84
    rectitis, 83
    retroversion, 87
    urethritis, 78
    phthisis, 76
    vaginitis, 78
Condition of the uterus that modifies the use of the pessary, 223
Conditions of the uterus in abortion, 59
Congestions, local, 140
Considerations, general, 17
Constipation, 62, 141
Contraction of the os uteri, 180
Convulsions, syncopal, 35
Corpus uteri, 108
Counter-irritant, seton as a, 198
Cramping pain, 54
Cures, spontaneous, 127
Cystitis as a complication, 124

Davidson's syringe, 158
Decompositions of productions of labor, 65
Deficient involution, treatment of 218
Depression or lapse, 210

Derangement, moral and mental, 41
Diagnosis, 101
  of cellulitis, 81
  between cancer and chronic inflammation of the cervix, 125
  of rectitis, 83
  of endocervicitis, 121
  of submucous inflammation, 122
Diagnostic effect of caustic applications, 122
Diet, 133
Digital examination, 103
  through the rectum, 108
Discharge, 170
Disease, uterine, sympathetic accompaniments of, 26
  of the bowels, sympathetic, 27
  skin, 76
  throat, 76
Disk, pessary, 222
Displacements, bearing down not caused by, 52
  causes of, 214
  of the uterus, 85
  symptoms of, 216
  their philosophy and treatment, 208
  nature of, 209
  theory of, 87
Distended bladder, 216
Duration of the flow, 56
Dysmenorrhœa, flexions of uterus most frequent causes of, 55

Effect of inflammation upon labor, 60
Effects of temporary excitement, 49
  of ulceration and inflammation on the functions of the uterus, 54
  of partial closure of os uteri on menstruation, 55
  of inflammation upon the post-partum condition, 61
Ellis, Dr., treatment of, 193
Endocervicitis, 94
  in the aged, 95
  in the virgin, 95
Etiology, 62
Examination, physical, of female genitals, 101
  anæsthetics useful in, 103
  use of chair in, 103
  table in, 103
  of urethral canal, 109
  digital, 103
Examinations, Dr. Sims's method of making, 117
Excitement, temporary, effects of, 49
Excitability, nervous, 137
Excretory organs, sympathy of, 39
Extension of inflammation to bladder and rectum, 33

INDEX. 245

Extent of cellulitis, 81
External inflammation combined with internal in childbearing women, 95

Favorable cases, how and when does relief come in, 73
Fistula in ano, 84
Flexible caustic-holder, 172
Flexions of uterus, most frequent causes of dysmenorrhœa, 55
　common complications of chronic inflammation, 88
Flow, duration of, 56
　manner of modified by inflammation, 55
Forceps, tenaculum, Nott's, 118
Forms of ulceration, 97
Fountain syringe, 159
　mode of using, 160
Function of generation affected by inflammation, 57
Functions of the uterus, effects of inflammation on, 54
　will they be restored, 73

General circumstances, change of, only temporary in their effects, 128
　considerations, 17
　symptoms, 134
　treatment, 127
　　main objects of, 134
Globe pessary, 222
Gonorrhœa, 65

Hardness with atrophy, 92
　and enlargement, treatment of, 199
Hemorrhage, remedy for, 186
Hemorrhoids, 63, 85
Hodge's pessary, 226
How is the pain produced, 52
How and when does relief come in favorable cases, 73
How to find the os uteri, 116
How long will it take to cure the inflammation, 72
Hydrate of chloral, use of in sleeplessness and pain, 139
Hyperæsthesia, 34
Hypertrophy, 92
　of the rectal mucous membrane, 85

Iliac region, pain in the, 48
　soreness, in the, 48
Ilium, pain in the side, above the, 48
Improper reading, 62

Inability to stand, 47
　to walk, 47
Indigestion, 153
Induration and enlargement of uterus, 90
Indulgence, sexual, 62
Inflammation, acute, 129
　acute, after parturition or abortion, 129
　acute, supervention of, 129
　aphthous, 100
　atrophy, as the result of, 124
　can it always be removed, 71
　characteristic signs of, 121
　complication of mucous with submucous, 99, 123
　effect upon labor, 60
　effects upon the post-partum condition, 61
　extension of to bladder and rectum, 33
　external, combined with internal in childbearing women, 95
　how long will it take to cure, 72
　in submucous tissue, progress of, 97
　intensity of, 96
　local, how long will it take to cure, 72
　manner of the flow modified by, 56
　mucous, 93
　　progress of, 97
　of cervix, complications of, 78
　position of, 89
　seat of, 93
　submucous, or fibro-cellular, 89
　　treatment of, 195
　urethral and cystic, 79
　uterine, kind of pain attendant upon, 54
Inflammatory ulceration, yellow leucorrhœa when there is, 51
Injections, 157
　accident in using, 163
　astringent, 161
　frequency of using, 161
　manner of using, 158
　medicated, 159
　quantity, 159
　should they be used in pregnancy, 165
　temperature of, 162
Instruments for using nitrate of silver, 171
Intensity of mucous inflammation, 96
Intercourse, sexual, 134
Intra-uterine injection, use of condemned, 157
Involution, imperfect, 215
　arrest of, 215
　deficient, treatment of, 218

INDEX.

Kind of displacements to which the pessary is adapted, 224
of mechanical support, 219
of pain attendant upon uterine inflammation, 54
of syringe, 158

Labor, 64
decomposition of productions of, 65
effect of inflammation upon, 60
Lapse or descent of uterus, 210
Leeching, 197
Length of the cervical and uterine cavities, 111
Lent's ointment syringe, 189
Leucorrhœa, 49
amount of, not always proportioned to extent of disease, 51
ulceration sometimes exists without, 51
yellow, when there is inflammatory ulceration, 51
yellow color of, always sign of ulceration, 51
Liver, sympathetic affection of, 28
Loaded intestinal canal, 216
Local congestions, 140
inflammation, how long will it take to cure, 72
symptoms, 47
treatment, 154
principles of, 129
that should govern in choosing the kind of, 167
Loins, pain in the, 47
Loss of a piece of nitrate of silver in the cervix, 188

Mammary bodies, 40
Manner of flow modified by inflammation, 56
Mechanical support, 219
Medicated injections, 159
Menorrhagia, 56
frequent in cervicitis, 57
Menstruation, pain during, 54
Mercury, acid nitrate of, 183
Mode of using the speculum, 115
Moral and mental derangement, 41
and intellectual perverseness, accompanying manifestations of, 34
Mucous inflammation, intensity of, 96
inflammation, progress of, 97
membrane, color of, 120
rectal, hypertrophy of the, 85
inflammation, 93
Mucus, indication of in abundance, 120
in the vagina, character of, 50

Muscular weakness, 36

Natural position of pelvic organs, 105
Nature of displacements, 209
Nervous system, sympathetic affection of the, 29
prostration, 135
excitability, 137
Nitrate of silver and its substitutes, 169
acid solution of, 176
effects of, 179, 180
form of application, 170
loss of a piece in the cervix, 188
mode of applying, 173, 174
instruments for using, 171, 176
is its application allowable in pregnancy, 187
kind of pain produced by application of, 178
results of extended experience with, 176
sometimes does harm, 183
sometimes fails, 182, 183
Nott's tenaculum forceps, 118

Object in using probe, 109
Ointment syringe, Lent's, 189
injecting, 188
Ointments, introduction of, 165
Organs, excretory, sympathy of the, 39
Os and cervix, appearance of, from fibrous induration, 91
in the aged, 119
in the multipara, 119
in the virgin, 118
Os uteri, contraction of the, 180
dilatation of the, 180
how to find, 116
in the aged, 107
partial closure of, 55
Oval pessary, 222
Ovaria, prolapse of, 217

Pain, cramping, 54
during menstruation, 54
in the iliac region, 48
in the loins, 47
in the sacral or lumbar region, 47
in the side, above the ilium, 48
how produced, 52
kind of, attendant upon uterine inflammation, 54
Pains, sympathetic, in the pelvic region, 32
Partial closure of os uteri, effects on menstruation, 55
Patient, position of during examination, 102

INDEX. 247

Pelvic organs, natural position of, 105
Perverseness, moral and intellectual, manifestations of, 34
Pessary, disk, 222
   globe, 222
   Hodge's, 226
   oval, 222
   ring, 221
   stem, 221
Pessaries, abdominal supporters, &c., 62, 221
Plethora, 140
Position of patient during examination, 102, 115
   of inflammation, 89
Posture, exercise and repose, 130
Post-partum condition, effects of inflammation upon, 61
Potash, caustic, object in using, 204
   mode of applying, 204
Pregnancy, 63
Pressure of the uterus upon the bladder, 87
Principles of local treatment, 129
Probe, mode of using, 110
   object in using, 109
   size and length of, 109
   and speculum conjointly, 120
Prognosis, 68
   in different varieties, 70
   influenced by age of patient, 72
   under treatment, 71
   without treatment, 69
Progress of inflammation in submucous tissue, 96
   of mucous inflammation, 97
   and terminations, 96
Prolapse, 211
   of the rectum, 84
   of uterus, 87, 211
   of ovaria, 217
Prostration, nervous, 135
Protrusion of uterus, 212
Pus, indication from, 120

Reading improper books, 62
Rectal mucous membrane, hypertrophy of, 85
Rectitis as a complication of uterine diseases, 83
   diagnosis of, 83
Rectum, stricture of, 83
   prolapse of, 84
Region, sacral or lumbar, pain in the, 47
   iliac, pain in the, 48
Remedy for the hemorrhage arising from the application of the nitrate of silver, 186
Remedies for the pain, 187

Remedies for nervousness, 187
Removal of tumors, 218
Respiration, 38
Retroflexion of uterus, 88
Retroversion, 87, 212
Ring pessary, 211

Sacral or lumbar region, pain in the, 47
Scarification, mode of doing, 172
Sciatic and anterior crural nerves, affections of the, 33
Seat of mucous inflammation, 93
Secretion, appearance of, 120
Senile uterus, 108
Seton as a counter-irritant, 196
Severe exertion, jolts, &c., 63
Severity of suffering not commensurate with amount of disease, 53
Sexual indulgence, 62
   intercourse, 134
Shower baths, 157
Sims's depressor, 117
   method of examining the uterus, 119
Size and length of probe, 109
   of uterus, 123
Skene's articulated uterine sound and knife, 199
Skin disease, 76
Slippery elm tent, 190
   introduced, 191, 192
Soreness in the iliac region, 48
Spasms, 35
Speculum, 113
   mode of using the, 115
   Emmet's, 118
   glass, 115
   quadrivalve, 114
   Storer's, 114
Spinal cord, affections of, 32
Sponge baths, 157
   tent, 201
Spontaneous cures, 127
Standing, 62
Sterility, 58
Stomach, sympathy of the, 26
Stricture of the rectum, 83
Submucous or fibro-cellular inflammation, 89
   inflammation, with ulceration and mucous inflammation, 195
   tissue, progress of inflammation in, 96
Subsidence of the uterus, 86
Suffering, severity of not commensurate with amount of disease, 53
Supervention of acute inflammation, 129
Support, mechanical, 219
Supporter, abdominal, in disease of cervix, 219

INDEX.

Supporter and pessary combined, 221
Supporters, pessaries, &c., abdominal, 62
Suppository syringe, 166
Suppuration in cellular tissue, 81
Swab for applying solutions, 177
Sympathetic accompaniments of uterine diseases, 26
   affection of the liver, 28
   affection of the nervous system, 29
   disease of the bowels, 27
   pains in the pelvic region, 32
Sympathy of the stomach, 26
   of the excretory organs, 39
Symptoms, local, 47
   of displacement, 216
Syncopal convulsions, 35
Syringe, kind of, 158
   Davidson's, 158
   hard-rubber, 171
   suppository, 166
System, circulatory, 36
   nervous, sympathetic affection of, 29

Temperature of baths, 156
   of injections, 162
Temporary excitement, effects of, 49
Tenaculum forceps, Nott's, 118
Tender uterus is an inflamed uterus, 108
Tenesmus, uterine, 49
Tents, or bougies, 189
   sponge, 192
Theory of displacements, 87
Throat disease, 76
Tissue, submucous, progress of inflammation in, 96
Treatment, general, 127
   main objects of, 134
   local, principles of, 129
   of deficient involution, 218
   of displacements, and their philosophy, 208, 218
   of Dr. Ellis, 193
   of hardness and enlargement, 199
   induration, 200
   of submucous inflammation, 195
Tumors, 216
   removal of, 218

Ulceration sometimes exists without leucorrhœa, 51
   forms of, 97
   and enlargement, 99
   inflammatory, yellow leucorrhœa when there is, 51
   yellow color, always sign of, 51

Urethral canal, examination of, 109
   and cystic inflammation, 79
Urethritis, 79
Uterine disease, sympathetic accompaniments of, 26
   and spermatorrhœa, comparison of symptoms of, 22
   inflammation, kind of pain attendant upon, 54
   probe or sound, 109
   mode of using, 110
   scarificator, 196
   sound and knife, Skene's, 199
   tenesmus, or weight, or bearing-down pain, 49
Uterometer, the, 111
   method of applying, for measuring the thickness of the uterus, 112
Uterus, anteversion of, 214
   conditions of in abortion, 59
   displacements of, 85
   effects of inflammation on the functions of the, 54
   general engorgement of, 91
   induration and enlargement of, 90
   of a childbearing woman, 107
   prolapse of, 87
   retroversion of, 213
   protrusion of, 212
   senile, 108
   size of, 123
   tender is an inflamed uterus, 108
Uterus and vagina, virgin, 106

Vagina, character of the mucus in the, 50
   decomposing substances in the, 65
Vaginitis, 65, 78
   of children, 66
Virgin, appearance of the os and cervix uteri in the, 118
   uterus and vagina, 106
   cervix, feel of, 104
   endocervicitis in, 95

Weakness, muscular, 36
Weight, or bearing-down pain, or uterine tenesmus, 49
Will the functions be restored, 73
Will the several symptoms always subside, 71

Yellow color always sign of ulceration, 51
   leucorrhœa, when there is inflammatory ulceration, 51

September, 1870.

CATALOGUE

OF

WORKS ON MEDICINE, SURGERY, DENTISTRY, AND THE COLLATERAL SCIENCES.

PUBLISHED BY

LINDSAY & BLAKISTON, PHILADELPHIA.

CONTENTS.

LINDSAY & BLAKISTON'S VISITING LIST FOR 1871, see reduction in prices.
TANNER'S PRACTICE OF MEDICINE, 5th American Edition (Just Ready.) Reprinted from the 6th Enlarged and Improved London Edition.
MEIGS AND PEPPER'S PRACTICAL TREATISE ON DISEASES OF CHILDREN, the Fourth Edition, very much enlarged.
CAZEAUX'S GREAT WORK ON OBSTETRICS, the most complete Text-Book on the subject now published.
AITKEN'S PRACTICE, from the 5th enlarged London Edition, with additions by Dr. Clymer equal to 500 pages of the English copy, made with special reference to the wants of the American Practitioner.
TROUSSEAU'S CLINICAL MEDICINE, 3 volumes (the third volume just ready.) Price reduced.
WORKS BY LIONEL S. BEALE, F.R.S., Prices reduced.
NEW BOOKS JUST READY AND IN PREPARATION.
ALPHABETICAL LIST of Lindsay & Blakiston's Medical Publications, with brief Critical or Descriptive Notices.
DENTAL WORKS published by Lindsay & Blakiston.
SCIENTIFIC BOOKS   do.   do.   do.
THE NEW SYDENHAM SOCIETY'S PUBLICATIONS, a Revised List, with announcements for 1870.
AMERICAN AND BRITISH PERIODICALS supplied by Lindsay & Blakiston.
ANATOMICAL MAPS AND PLATES.
ENGLISH BOOKS RECENTLY IMPORTED.
CONDENSED LIST OF LINDSAY & BLAKISTON'S PUBLICATIONS.

LINDSAY & BLAKISTON, in issuing a revised Catalogue of their Medical Publications, desire to call the attention of the Profession to the reduction they have made in the price of their **Visiting List for 1871**. This little work having become an almost indispensable companion to the Practising Physician, they wish to extend as far as possible its usefulness by keeping the price down as low as the increased cost of manufacture will permit. They have also, in order to meet to some extent the demand for lower prices, reduced the price of Dr. BEALE's books of TROUSSEAU'S CLINICAL MEDICINE, the 3d volume of which has just been issued, and of many other books in their Catalogue, and also their rates for the importation of FOREIGN BOOKS and PERIODICALS.

☞ ORDERS for any MEDICAL BOOKS, whether published in the United States or otherwise, and not contained in this Catalogue, will be executed promptly, and at the lowest prices.

PRICES REDUCED OF
LINDSAY & BLAKISTON'S
PHYSICIAN'S VISITING LIST.
NOW READY FOR 1871.

"The simplest of all the VISITING LISTS published, it must continue to hold, what it now has, the preference over all other forms of this indispensable companion for the Physician."—*New York Med. Journal.*

CONTENTS.

1. Table of Signs, or Guide for Registering Visits, Engagements, &c.
2. An Almanac for 1868.
3. Marshall Hall's Ready Method in Asphyxia.
4. Poisons and their Antidotes.
5. Table for Calculating the Period of Utero-Gestation.
6. The Visiting List arranged for 25, 50, 75, or 100 Patients.
7. Memoranda pages for every month in the year.
8. Pages for Addresses of Patients, &c.
9. " Bills and Accounts asked for and delivered.
10. " Obstetric Engagements.
11. " Vaccination.
12. " Recording Obstetric Cases, Deaths, and for General Memoranda.

SIZES AND PRICE.

For 25 Patients weekly.	Tucks, pockets, and pencil,	$1 00
50 " " " " " "		1 25
75 " " " " " "		1 50
100 " " " " " "		2 00
50 " "2 vols. {Jan. to June. / July to Dec.} "		2 50
100 " "2 vols. {Jan. to June. / July to Dec.} "		3 00

Also, AN INTERLEAVED EDITION,

for the use of *Country Physicians* and others who compound their own Prescriptions, or furnish *Medicines* to their patients. The additional pages can also be used for *Special Memoranda*, recording important cases, &c., &c.

For 25 Patients weekly, interleaved, tucks, pockets, etc.,		$1 50
50 " " " " " "		1 75
50 " " 2 vols {Jan. to June. / July to Dec.} " "		3 00

This VISITING LIST has now been published for *Twenty Years*, and has met with such uniform and hearty approval from the Profession, that the demand for it has steadily increased from year to year.

The Publishers, in order to still further extend its circulation and usefulness, and to keep up the reputation which it has so long retained, of being

THE CHEAPEST AND BEST,

as well as the OLDEST VISITING LIST published, *have now made a very considerable reduction in the price.*

It can be procured from the *principal booksellers* in any of the *large cities* of the *United States and Canada*, or copies will be forwarded by mail, *free of postage*, by the Publishers, upon receipt by them of the retail price as annexed.

In ordering the work from other booksellers, order

Lindsay & Blakiston's Physician's Visiting List.

And in all cases, whether ordering from the Publishers or otherwise, specify the size, style, &c., wanted.

LINDSAY & BLAKISTON, Publishers,
25 South Sixth St., Philadelphia.

"The leading feature of this book is its essentially practical character."—
LONDON LANCET.

Tanner's Practice of Medicine.

FIFTH AMERICAN, FROM THE SIXTH LONDON EDITION.

ENLARGED AND THOROUGHLY REVISED.

JUST READY.

THE PRACTICE OF MEDICINE, by THOMAS HAWKES TANNER, M.D., *Fellow of the Royal College of Physicians, Author of Tanner's Practical Treatise on the Diseases of Children, &c., &c. Fifth American Edition, with a very large Collection of Formulæ. One Volume, Royal Octavo, containing over 1100 pages.*

Price, handsomely bound in Cloth, . . $6 00
" " " Leather, . . 7 00

CONTENTS.

Part 1. General Diseases.
" 2. Fevers.
" 3. Venereal Diseases.
" 4. Diseases of the Nervous System.
" 5. Diseases of the Organs of Respiration and Circulation.
" 6. Diseases of the Thoracic Walls.
" 7. Diseases of the Alimentary Canal.
" 8. Diseases of the Liver.
" 9. Diseases of the Pancreas and Spleen.

Part 10. Diseases of the Abdominal Walls.
" 11. Diseases of the Urinary Organs.
" 12. Diseases of the Uterine Organs.
" 13. Diseases of the Skin.
" 14. Diseases of Cutaneous Appendages.
" 15. Diseases of the Bloodvessels.
" 16. Diseases of the Absorbent System.
Appendix of Formulæ
General Index.

"The rapidity with which edition after edition of this work has appeared and disappeared is, on the whole, a true test of its merits. The fifth edition was, we believe, a very large one, yet the book was for some time out of print before the present one could be prepared. Dr. Tanner has chosen his title well; his work is essentially one on the practice of medicine in its widest sense, and it is in what relates to pure practice, as contradistinguished from the theory of medicine, that the book is strongest; for it has been the author's aim to collect everything he could think of which would aid the practitioner in the discharge of his duties. But it is not to men engaged in the active discharge of the duties of their profession alone to whom the book is welcome. With the student, preparing himself to enter upon these duties, the book has long been a favorite, chiefly, we believe, from the lucidity of its style and the character of its substance. Other books there are, more eloquent and more recondite, but none excel Dr. Tanner's work in these important features. All that is necessary to know is here, disposed in such a manner as to admit of the readiest reference, and of being most easily retained in the memory. Our limits will not admit of an extended review, which would be out of place with regard to a book practically established as a standard. It carries its own recommendation, and is its own best passport to general use. It has been the result of very great labor—labor well spent; and it appears in a form which is creditable to its publishers as it is pleasing to those who have to use the book.—*British and Foreign Medico-Chirurgical Review*, April, 1870.

"Dr. Tanner's works are all essentially and thoroughly practical,—he never for one moment allows this utilitarian end to escape his mental view. He aims at teaching how to recognize and how to cure disease, and in this he is thoroughly successful. It is indeed a wonderful mine of knowledge."—*Medical Times and Gazette*, July, 1869.

MEIGS AND PEPPER ON CHILDREN.

"*The most thorough and Practical Work on the subject now before the Profession.*"

Meigs and Pepper's Practical Treatise on the Diseases of Children.

Fourth Edition, thoroughly Revised and greatly Enlarged.

By J. FORSYTH MEIGS, M. D., *Fellow of the College of Physicians of Philadelphia, &c., &c.*, and WILLIAM PEPPER, M. D., *Physician to the Philadelphia Hospital, &c., &c.*, forming a Royal Octavo Volume of over 900 pages. Price, bound in Cloth, . . . $6 00
" " Leather, . . . 7 00

Dr. Meigs' work has been out of print for some years. The rapid sale of the three previous editions, and the great demand for a new edition, is sufficient evidence of its great popularity; while the very large practice of many years' standing of the author in the speciality of "Diseases of Children," imparts to it a value unequalled, probably, by any other work on the same subject now before the Profession. This present edition has been almost entirely rewritten and rearranged, and no effort or labor has been spared by either Drs. Meigs or Pepper, to make it represent fully in its most advanced state the present condition of Medicine as applied to Children's Diseases.

The entire work has been subjected to careful revision. Several of the articles, as those on Eclampsia, Chorea, and Parasitic Skin Diseases, have been much enlarged; and others, as the various articles on the Diseases of the Stomach and Intestines, and that on Eczematous Affections, entirely rewritten. In addition, articles have been added upon the following important subjects:

Diseases of the Heart.
Cyanosis.
Diseases of the Cæcum and Appendix.
Intussusception.
Chronic Hydrocephalus.
Tetanus Nascentium.
Atrophic Infantile Paralysis.
Progressive Paralysis, with apparent Hypertrophy of the Muscles.

Facial Paralysis.
Rheumatism.
Diphtheria.
Mumps.
Rickets.
Tuberculosis.
Infantile Syphilis.
Typhoid Fever.
Scleroma.

The new matter thus added amounts to nearly 200 pages. It has been the effort of the authors, while endeavoring to make the work fully represent the state of our knowledge upon the subjects treated of, to retain its eminently practical character; and with this view, an unusually large amount of space has been devoted to the consideration of the treatment of each disease.

"This is the fourth edition of Meigs on Diseases of Children, greatly enlarged and improved by chapters upon a large number of new subjects, and also by a very copious index, which facilitates reference, and makes the work more serviceable to the Practitioner. As now enlarged, it is one of the most complete and comprehensive works of its class, and will meet the wants of the Profession in this department most admirably.'—*Buffalo Med. and Surg. Journal.*

"It is very comprehensive, and embraces most of the maladies incident to childhood and infancy. We consider it a very safe, reliable, and suggestive guide, being quite large and full in detail, embracing almost everything pertaining to the subject, making it a very useful book both for reference and study."—*Medical Archives.*

"It forms the most complete and comprehensive work upon the diseases of children published in this country. It has for years been one of the standard authorities, and in its present enlarged form will still more command attention. It presents the latest views of pathology and treatment, and takes into consideration many subjects which were entirely omitted in the previous editions."—*Detroit Journal of Medicine, &c.*

"It is satisfactory to note that the authors have brought up their work to the level of the pathological knowledge of the day, and that their therapeutical notions are equally advanced. The authors are enrolled among the more enlightened therapeutists of our time. One cannot fail to be struck throughout the treatise with the very judicious advice given by the authors on various points of treatment. The work, as a whole, is entitled to rank with the best."—*Medical Repertory.*

Cazeaux's Great Work on Obstetrics.

THE MOST COMPLETE TEXT-BOOK NOW PUBLISHED.
GREATLY ENLARGED AND IMPROVED.
CONTAINING 175 ILLUSTRATIONS.

A Theoretical and Practical Treatise on Midwifery, *including the Diseases of Pregnancy and Parturition, by* P. CAZEAUX, *Member of the Imperial Academy of Medicine; Adjunct Professor in the Faculty of Medicine of Paris, etc., etc. Revised and Annotated by* S. TARNIER, *Adjunct Professor in the Faculty of Medicine of Paris; Former Clinical Chief of the Lying-in-Hospital, etc., etc. Fifth American from the Seventh French Edition. Translated by* WM. R. BULLOCK, M. D. In one volume Royal Octavo, of over 1100 pages, with numerous Lithographic and other Illustrations on Wood.

Price, bound in Cloth, bevelled boards, . . $6.50
" " Leather, ... 7.50

M. Cazeaux's Great Work on Obstetrics has become classical in its character, and almost an Encyclopædia in its fulness. Written expressly for the use of students of medicine, and those of midwifery especially, its teachings are plain and explicit, presenting a condensed summary of the leading principles established by the masters of the obstetric art, and such clear, practical directions for the management of the pregnant, parturient, and puerperal states, as have been sanctioned by the most authoritative practitioners, and confirmed by the author's own experience. Collecting his materials from the writings of the entire body of antecedent writers, carefully testing their correctness and value by his own daily experience, and rejecting all such as were falsified by the numerous cases brought under his own immediate observation, he has formed out of them a body of doctrine, and a system of practical rules, which he illustrates and enforces in the clearest and most simple manner possible.

OPINIONS OF THE PRESS.

"It is unquestionably a work of the highest excellence, rich in information, and perhaps fuller in details than any text-book with which we are acquainted. The author has not merely treated of every question which relates to the business of parturition, but he has done so with judgment and ability."—*British and Foreign Medico-Chirurgical Review.*

"The translation of Dr. Bullock is remarkably well done. We can recommend this work to those especially interested in the subjects treated, and can especially recommend the American edition."—*Medical Times and Gazette.*

"The edition before us is one of unquestionable excellence. Every portion of it has undergone a thorough revision, and no little modification; while copious and important additions have been made to nearly every part of it. It is well and beautifully illustrated by numerous wood and lithographic engravings, and, in typographical execution, will bear a favorable comparison with other works of the same class."—*American Medical Journal.*

"In the multitudinous collection of works devoted to the propagation of human beings, and to the details of parturition, none, in our estimation, bears any comparison to the work of Cazeaux, in its entire perfectness; and if we were called upon to rely alone on one work on accouchments, our choice would fall upon the book before us without any kind of hesitation."—*West. Jour. of Med. and Surgery.*

"We do not hesitate to say, that it is now the most complete and best treatise on the subject in the English language."—*Buffalo Medical Journal.*

"We know of no work on this all-important branch of our profession that we can recommend to the student or practitioner as a safe guide before this."—*Chicago Medical Journal.*

"Among the many valuable treatises on the science and art of obstetrics, the work of Cazeaux stands pre-eminent."—*St. Louis Med. and Surg. Journal.*

"M. Cazeaux's book is the most complete we have ever seen upon the subject. It is well translated, and reflects great credit upon Dr. Bullock's intelligence and industry."—*N. A. Medico-Chirurg. Review.*

"The Representative Book of Medical Science."—LONDON LANCET.

Aitken's Science and Practice of Medicine.
SECOND AMERICAN FROM THE FIFTH LONDON EDITION.
Containing ADDITIONS equal to 500 pages of the English Edition, prepared by the AMERICAN EDITOR with Special Reference to the wants of the AMERICAN PRACTITIONER.

The Science and Practice of Medicine. By WILLIAM AITKEN, M.D. *Second American from the Fifth Enlarged London Edition, with Additions by* MEREDITH CLYMER, M.D., *late Professor of the Institutes and Practice of Medicine in the University of New York, Physician to the Philadelphia Hospital, etc.* In 2 Volumes Royal Octavo.

With a COLORED MAP, a LITHOGRAPHIC PLATE, and ONE HUNDRED AND THIRTY ILLUSTRATIONS ON WOOD.

Price, bound in Cloth, bevelled boards $12.00
" " Leather, ... 14.00

Fifteen months have been spent by Dr. AITKEN in thoroughly revising this *Great Work*, and adding to it many valuable additions and improvements amounting to about 100 pages of new matter, included in which will be found the adoption and incorporation in the text of the *"new nomenclature of the Royal College of Physicians of London;"* to which are added the *Definitions* and the Foreign equivalents for their English names.

The subjects of *Malignant Cholera*, of *Paralysis*, of *Epidemic Cerebro-Spinal Meningitis*, and of *Intestinal Obstruction* have been entirely re-written; and several other subjects in connection with the treatment of disease, of the greatest importance, are considered for the first time in this edition.

The Press have referred to former editions of this work as "an admirable compilation." "The most comprehensive of any in the practice of medicine." "It embodies the most advanced knowledge of the time." "The most valuable class-book for students yet published." "It may be looked upon as the standard text-book in the English language." "The present work contains information that will not be found in any other Manual of Medicine," &c., &c.

The author in this edition has endeavored to keep up this high reputation, and to make it in every respect *a Representative Book of Medical Science and the Practice of the day*, as understood and followed by the best men of the Profession.

Large additions have also been made by the American Editor, Dr. MERIDITH CLYMER, equal to over 500 pages of the London edition, and with particular reference to the wants of the AMERICAN PRACTITIONER, included in which are new articles upon the following subjects: *Spinal Symptoms of Typhoid Fever, Typho-Malarial Fever, Chronic Camp Dysentery, Cholera Morbus, Cholera Infantum, Hereditary Syphilis, Gonorrhœal Rheumatism, The Delirium of Inanition, Chronic Alcoholism, Syphilitic Disease of the Liver, Epileptiform Neuralgia, Capillary Bronchitis, Plastic Bronchitis, Dilatation of the Bronchia, Fibroid Degeneration of the Lung, Chronic Pyæmia, &c. &c.*

Trousseau's Clinical Lectures.

VOL. III. NOW READY.

Lectures on Clinical Medicine, delivered at the Hôtel-dieu, Paris. By A. TROUSSEAU, *Professor of Clinical Medicine in the Faculty of Medicine, Paris, &c., &c.*

Trousseau's Lectures on Clinical Medicine, so favorably received, as well by the profession of the United States as abroad, are published in this country in connection with the New Sydenham Society, under whose auspices the translation of Vols. II. and III. have been made. Either of these volumes can be furnished separately, and in order to still further extend the circulation of so valuable a work, the Publishers have now reduced the price to *Five Dollars* per volume.

CONTENTS OF VOLUME I.—*Translated and Edited by P. Victor Bazire, M. D., &c.*—LECTURE 1. On Venesection in Cerebral Hæmorrhage and Apoplexy. 2. On Apoplectiform Cerebral Congestion, and its Relations to Epilepsy and Eclampsia. 3. On Epilepsy. 4. On Epileptiform Neuralgia. 5. On Glosso-laryngeal Paralysis. 6. Progressive Locomotor Ataxy. 7. On Aphasia. 8. Progressive Muscular Atrophy. 9. Facial Paralysis, or Bell's Paralysis. 10. Cross-paralysis, or Alternate Hemiplegia. 11. Infantile Convulsions. 12. Eclampsia of Pregnant and Parturient Women. 13. On Tetany. 14. On Chorea. 15. Senile Trembling and Paralysis Agitans. 16. Cerebral Fever. 17. On Neuralgia. 18. Cerebral Rheumatism. 19. Exophthalmic Goitre, or Graves' Disease. 20. Angina Pectoris. 21. Asthma. 22. Hooping Cough. 23. On Hydrophobia.

CONTENTS OF VOLUME II.—*Translated from the Edition of 1868 (being the last revised and enlarged edition). by John Rose Cormack, M. D., Edin., F.R.S.E , &c.*—LECTURE 1. Small-pox. 2. Variolous Inoculation. 3. Cow-pox. 4. Chicken-pox. 5. Scarlatina. 6. Measles, and in particular its unfavorable Symptoms and Complications. 7. Rubeola. 8. Erythema Nodosum. 9. Erythema Papulatum. 10. Erysipelas, and in particular Erysipelas of the Face. 11. Mumps. 12. Urticaria. 13. Zona, or Herpes Zoster. 14. Sudoral Exanthemata. 15. Dothinenteria, or Typhoid Fever. 16. Typhus. 17. Membranous Sore Throat, and in particular Herpes of the Pharynx. 18. Gangrenous Sore Throat. 19. Inflammatory Sore Throat. 20. Diphtheria. 21. Thrush.

CONTENTS OF VOLUME III.—*Translated from the Edition of 1868. by John Rose Cormack, M.D., Edin., F.R.S E., &c.*—LECTURE 22. Specific Element in Disease. 23. Contagion. 24. Ozæna. 25. Stridulous Laryngitis, or False Croup. 26 Œdema of the Larynx. 27. Aphonia: Cauterization of the Larynx. 28. Dilatation of the Bronchi and Bronchorrhœa. 29. Hemoptysis. 30. Pulmonary Phthisis. 31. Gangrene of the Lung. 32. Pleurisy: Paracentesis of the Chest. 33. Traumatic Effusion of Blood into the Pleura: Paracentesis of the Chest. 34. Hydatids of the Lung. 35. Pulmonary Abscesses and Peripneumonic Vomicæ. 36. Treatment of Pneumonia. 37. Paracentesis of the Pericardium. 38. Organic Affections of the Heart. 52. Alcoholism. 62. Spermatorrhœa. 63. Nocturnal Incontinence of Urine. 64. Glucosuria: Saccharine Diabetes. 65. Polydipsia. 67. Vertigo a Stomacho Læso.

3 Volumes Octavo, handsomely bound in Cloth, Price $5 00 each.

OPINIONS OF THE PRESS.

"Trousseau furnishes us with an example of the best kind of Clinical teaching. It is a book that deserves to be popularized. The translation is perfect." — *Medical Times and Gazette.*

"The great reputation of Prof. Trousseau as a practitioner and teacher of Medicine in all its branches, renders the present appearance of his Clinical Lectures particularly welcome." — *Medical Press and Circular.*

"The publication of Trousseau's Lectures will furnish us with one of the very best practical treatises on disease as seen at the bedside." — *British and Foreign Medico-Chirurgical Review.*

"A clever translation of Prof. Trousseau's admirable and exhaustive work, the best book of reference upon the Practice of Medicine." — *Indian Medical Gazette.*

"The Lectures of Trousseau, in attractiveness of manner and richness of thoroughly practical matter, worthily takes a place beside the classical lectures of Watson and Graves." — *British Medical Journal.*

"Trousseau is essentially the French Graves, and his lectures should sooner than this have been translated into English." — *Lancet.*

WORKS BY DR. LIONEL S. BEALE, F.R.S.,

Professor of Physiology and of General and Morbid Anatomy in King's College, London, &c., &c.

PRICES REDUCED.

Beale's How to Work with the Microscope.

SEVENTH THOUSAND—400 ILLUSTRATIONS, *some of which are Colored, together with a Photographic Frontispiece.*

This work is a complete manual of microscopical manipulation, and contains a full description of many new processes of investigation, with directions for examining objects under the highest powers, and for taking photographs. It is got up, both plates and letter-press, in an elegant manner, and is, without doubt, the most complete and beautiful book of the kind in the English language.

Octavo, cloth. Price, ... $7 50

On Kidney Diseases, Urinary Deposits, and

CALCULOUS DISORDERS. *Including the Symptoms, Diagnosis, and Treatment of Urinary Diseases. With full Directions for the Chemical and Microscopical Analysis of the Urine in Health and Disease.* The Third Edition, very much enlarged.

This Work is Illustrated with 70 Plates, containing upwards of 400 separate figures, all of which have been copied from Nature, and many are now published for the first time.

A handsome volume Octavo. Bevelled, cloth. Price, . . $10 00

The complete and thoroughly practical character of this work may be judged of by the fact that it contains as many as *thirty-six figures* of different forms of Uric Acid alone.—The text and plates have been printed on toned paper, with the utmost care, and the work is the most complete, as well as the largest, ever published on the subject.

Beale on the Microscope in Practical Medicine

FIVE HUNDRED ILLUSTRATIONS.

The Microscope in its Application to Practical Medicine; by LIONEL S. BEALE, M.B., F.R.S., &c. The Fourth Thousand, thoroughly Revised by the Author, with nearly 100 pages of New Matter and numerous additional Illustrations. In One Volume Octavo. Price, . $7 00

Beale's Protoplasm;

Or, Life, Matter, and Mind. Second Edition, much Enlarged, with Eight Colored Illustrations; and an entirely new section on Mind By LIONEL S. BEALE, M.B., F.R.S, *and Author of "How to Work the Microscope," "The Microscope in Practical Medicine," &c., &c.* In One Volume. Price, ... $3 00

Beale on Disease Germs,

Their Supposed Nature. With Colored Plates, uniform with "PROTOPLASM." In One Volume 12mo. Cloth. Price, . . $1 7

NEW BOOKS. JUST READY.

Althaus's Medical Electricity.
A NEW AND VERY MUCH ENLARGED EDITION.

A Treatise on Medical Electricity, Theoretical and Practical, and its Use in the Treatment of Paralysis, Neuralgia, and other Diseases. The Second Edition, revised, and for the most part re-written. By JULIUS ALTHAUS, M.D., *Member of the Royal College of Physicians, Senior Physician to the Infirmary for Epilepsy and Paralysis, &c., &c.* In one volume, octavo. Illustrated by a Lithographic Plate and sixty-two Engravings on Wood. Price, ... $5 00

It is with great pleasure that we welcome, and cordially recommend, Dr. Althaus's treatise, in the improved guise of a Second Edition. Dr. Althaus's work possesses the merit of being clearly and agreeably written, while its teaching is in accord with the most recent information; and the balance is evenly held between the relative virtues of galvanization and faradization—a point of the highest importance in the treatment of disease.

To the favorable opinions already accorded to the former edition of this treatise we can add nothing, except in the way of sincere commendation; and to Dr. Althaus belongs the credit of having filled up a hiatus in English medical literature, by the production of a sound, comprehensive, and practical work on the right use of an agent whose powers are daily becoming better recognized and more clearly defined.—*Dublin Quarterly*, May, 1870.

Tyson's Work on the Cell Doctrine.

The Cell Doctrine: its History and Present State. With a Copious Bibliography of the Subject. For the Use of Students of Medicine and Dental Surgery. By JAMES TYSON, M.D., *Lecturer on Microscopy in the University of Pennsylvania, and on Physiology in the Pennsylvania College of Dental Surgery; Fellow of the College of Physicians of Philadelphia, &c., &c.* In one volume, with a Colored Plate, and numerous Illustrations on Wood. Price, . . . $2 00

Dr. Tyson furnishes in this work a concise and instructive *resumé* of the origin and advance of the doctrine of Cell evolution. In it we find the theories of Virchow, Robin, Huxley, Hughes, Bennett, Beale, and other distinguished men. Its pages contain what could otherwise only be learned by the perusal of many works, and supplies the reader with a continuous, complete, and general knowledge of the history, progress, and peculiar phases of the Cell Doctrine, accompanied by careful references and a copious bibliography.

Legg on Urine. FROM THE SECOND LONDON EDITION.

A Guide to the Examination of the Urine. For the Practitioner and Student. By J. WICKHAM LEGG, M.D., *Member of the Royal College of Physicians, &c., &c.* Second Edition. 16mo. Cloth. Price, 75 cts.

Dr. Legg's little manual has met with remarkable success, and the speedy exhaustion of the first edition has enabled the author to make certain emendations which have added greatly to its value. We can now confidently commend it to the student as a safe and reliable guide to such examinations of the urine as he may be called upon to make.—*London Medical Times and Gazette.*

Kirkes' Hand-Book of Physiology.
THE SEVENTH LONDON EDITION.

HAND-BOOK OF PHYSIOLOGY, by WILLIAM SENHOUSE KIRKES, M.D. Seventh Edition, Edited by W. MORRANT BAKER, F.R.C.S., *Lecturer on Physiology, &c., &c.* With 241 Illustrations. In one volume, demy-octavo, containing over 800 pages. Price, bound in cloth, $5 00

This edition of Dr. Kirkes' Hand-Book of Physiology is fully brought up to the times, and forms one of the most complete and convenient TEXT-BOOKS on the subject, for the Student of Medicine, now in print.

NEW BOOKS, JUST READY.

J. Sœlberg Wells on the Eye.

A NEW ENLARGED LONDON EDITION.

A Treatise on Diseases of the Eye (the Author's Second Edition), illustrated by Colored Plates and numerous Engravings on Wood. By J. SŒLBERG WELLS, *Professor of Ophthalmology in King's College, London, &c., &c.* The plates and letter-press elegantly printed on cream-tinted paper. Octavo, bound in Cloth. Price, . . $6 50
  Do. " Leather, " . . 7 50

THE AUTHOR'S PREFACE TO HIS SECOND EDITION.

It has afforded me no small gratification that the First Edition of this work should have met with so favorable a reception, both by the Profession at large and by the British and Foreign Medical Press, and especially that it should have been deemed worthy of being translated into French and German, in both of which languages it will be published in the course of this year.

Stimulated by such encouragement, I have endeavored to render the Second Edition as complete as possible, and have made numerous additions, incorporating all the important facts elucidated by the most recent researches, so that the work might be brought up to the latest date.—16 *Saville Row, London,* May, 1870.

Coles on Deformities of the Mouth.

SECOND EDITION, NOW READY.

On Deformities of the Mouth, Congenital and Acquired, with their Mechanical Treatment. By JAMES OAKLEY COLES, *Dentist to the Hospital for Diseases of the Throat, Member of the Odontological Society, &c., &c.* Second Edition, Revised and Enlarged, containing *Eight Colored Plates and Fifty-one Engravings on Wood.*
  One Volume, Demy-Octavo, in Cloth. Price, . . . $2 50

OPINIONS OF THE PRESS ON FIRST EDITION.

" This work is full of useful information. The records of cases treated are most encouraging."—*Lancet.*

" This beautifully illustrated work deals with a highly practical and interesting subject."—*Medical Times and Gazette.*

" We take pleasure in commending it as a valuable, and therefore desirable, treatise to possess."—*Dental Cosmos.*

Neumann's Hand-Book of Skin Diseases.

Translated from the Author's Second Edition. IN PREPARATION.

A Hand-Book of Skin Diseases, illustrated by Forty-nine Wood Cuts of large size, beautifully executed, and showing the Microscopic Appearance of Sections of the Skin in its Various Diseases. By PROF. NEUMANN, of Vienna. In one Volume.

Dr. Neumann was for a long time Prof. Hebra's assistant, and his work is a concise treatise founded on "Hebra's" Doctrines and Methods of Treatment, as the latter work may not for a long time be completed, and as from its size and its publication in the English language only by the New Sydenham Society, it cannot, even when all published, be accessible but to a limited number. This work presenting his views, must necessarily prove a great acquisition to the profession.

LINDSAY & BLAKISTON'S
MEDICAL PUBLICATIONS.

"The Representative Book of Medical Science."—LONDON LANCET.

Aitken's Science and Practice of Medicine.

SECOND AMERICAN from the FIFTH LONDON EDITION.

In 2 Volumes, containing 2000 Royal Octavo Pages, a COLORED MAP, a LITHOGRAPHIC PLATE, and ONE HUNDRED AND THIRTY ILLUSTRATIONS ON WOOD.

Price, bound in Cloth, bevelled boards $12.00
" " Leather 14 00

Dr. Aitken's work is now the most comprehensive TEXT-BOOK on the Practice of Medicine in the English Language; the present edition has been enlarged and carefully revised by the author, as well as by the American editor, MERIDITH CLYMER, M.D., the latter having made additions of original matter equal to nearly 500 pages of the English Edition, with special reference to the wants of the American Practitioner.

Acton on the Functions and Disorders of the Reproductive Organs.

Second American from the Fourth London Edition. Carefully revised by the author, with additions. Just Ready, octavo, . . . $3.00

"We think Mr. Acton has done good service to society by grappling manfully with sexual vice, and we trust that others, whose position as men of science and teachers enable them to speak with authority, will assist in combating and arresting the evils which it entails. We are of the opinion that the spirit which pervades it is one that does credit equally to the head and to the heart of the author."—*The British and Foreign Medico-Chirurgical Review.*

Anstie on Stimulants and Narcotics.

Their Mutual Relations, with Special Researches on the Action of Alcohol, Ether, and Chloroform on the Vital Organism. By FRANCIS E. ANSTIE, M.D., *Assistant Physician to Westminster Hospital, Lecturer on Materia Medica and Therapeutics, etc., etc.* Octavo, ... $3.00

Althaus' Medical Electricity. *A New and Very Much Enlarged Edition.*

A Theoretical and Practical Treatise, and its Use in the Treatment of Paralysis, Neuralgia, and other Diseases. By JULIUS ALTHAUS, M.D., *Member of the Royal College of Physicians, &c. Second Edition, revised, enlarged, and for the most part rewritten. In One Volume Octavo, with a Lithographic Plate and sixty-two Illustrations on Wood.* Price, ... $5.00

PROF. BYFORD'S NEW EDITION
OF

The Practice of Medicine and Surgery,

Applied to the Diseases and Accidents Incident to Women. By W. H. BYFORD, A. M., M. D., *Professor of Obstetrics and Diseases of Women and Children in the Chicago Medical College, &c., &c.* The Second Edition, revised and enlarged, with additional illustrations. One volume octavo.

Bound in Cloth. Price, ... $5.00
" Leather, ... 6.00

The rapid sale of the first edition of this book, which was exhausted in a little more than a year, has enabled the author to carefully revise the whole work, add many improvements, and to make a large addition of new matter, without, however, materially increasing the size of the volume.

"Professor Byford's book may safely be said to fill a gap in a most important department of the healing art. The treatise is as complete a one as the present state of our science will admit of being written, and not only gives us the cases in which the knife is to be called into requisition, but fully discusses all those uterine ailments which are to be benefited by medical treatment. In this latter respect, the work has a peculiar value to every general practitioner. The author speaks from experience, evidently with the sole object of presenting his subject in a strictly impartial and scientific light. The present edition is much enlarged in point of matter contained in the work, although the volume itself is very little, if any, increased in bulk. We commend it to the diligent study of every practitioner and student, as a work calculated to inculcate sound principles, and lead to enlightened practice."—*N. Y. Med. Record.*

"This work treats well-nigh all the diseases incident to women, diseases and accidents of the vulva and perineum, stone in the bladder, inflammation of the vagina, menstruation and its disorders, the uterus and its ailments, ovarian tumors, diseases of the mammæ, puerperal convulsions, phlegmasia alba dolens, puerperal fever, &c. Its scope is thus of the most extended character, yet the observations are concise, but convey much practical information."—*London Lancet.*

BYFORD ON THE UTERUS.—PREPARING.
A New, Enlarged, and Illustrated Edition.

On the Chronic Inflammation and Displacements of the Unimpregnated Uterus.

A New, Enlarged, and Thoroughly Revised Edition, with Illustrations. One Volume. Octavo.

Biddle's Materia Medica. The Third Edition Enlarged.

For the Use of Students. A New Enlarged Edition. With Illustrations. By J. B. BIDDLE, M. D., *Professor of Materia Medica and Therapeutics in the Jefferson Medical College, Philadelphia, &c., &c.* Octavo. $4.00

This is a thoroughly revised and enlarged edition of Prof. Biddle's work on Materia Medica. It is designed to present the leading facts and principles usually comprised under this head, as set forth by the standard authorities, and to fill a vacuum which seems to exist in the want of an elementary work on the subject. The larger works usually recommended as Text-Books in our Medical Schools are too voluminous for convenient use. This work will be found to contain, in a condensed form, all that is most valuable, and will supply students with a reliable guide to the courses of lectures on Materia Medica, as delivered at the various Medical Schools in the United States

Beasley's Book of Prescriptions.

A NEW, REVISED, AND ENLARGED EDITION.

Containing 3000 Prescriptions, collected from the Practice of the most Eminent Physicians and Surgeons — English, French, and American; comprising also a Compendious History of the Materia Medica, Lists of the Doses of all Officinal and Established Preparations, and an Index of Diseases and their Remedies. By HENRY BEASLEY. Second American from the last London Edition. Octavo. $4.00

"The editor, carefully selecting from the mass of materials at his disposal, has compiled a volume, in which both physician and druggist, prescriber and compounder, may find, under the head of each remedy, the manner in which that remedy may be most effectively administered, or combined with other medicines in the treatment of various diseases. The alphabetical arrangement adopted renders this easy; and the value of the volume is still further enhanced by the short account given of each medicine, and the lists of doses of its several preparations. It is really a most useful and important publication, and, from the great aid which it is capable of affording in prescribing, should be in the possession of every medical practitioner. Amongst other advantages is, that, by giving the prescriptions of some of the most able and successful practitioners of the day, it affords an insight into the methods of treatment pursued by them, and of the remedies which they chiefly employed in the treatment of different diseases." — *Lancet.*

Beasley's Druggists' General Receipt Book.

SIXTH AMERICAN EDITION, REVISED AND IMPROVED.

Comprising a copious Veterinary Formulary, numerous Receipts of Patent and Proprietary Medicines, Druggists' Nostrums, etc.; Perfumery and Cosmetics, Beverages, Dietetic Articles and Condiments, Trade Chemicals, Scientific Processes, and an Appendix of Useful Tables, by HENRY BEASLEY, Author of the Book of Prescriptions, etc., etc. Sixth American from the Last London Edition. Octavo. $3.50

"This is one of the class of books that is indispensable to every Druggist and Pharmaceutist as a book of reference for such information as is wanted, not contained in works used in the regular line of his business, and we can recommend it as one of the best of the kind." — *American Druggists' Circular.*

Barth & Roger's Manual of Auscultation and Percussion.

A new American Translation from the Sixth French Edition. 16mo. $1.25.

"This is one of the most useful and practical manuals of its sort that has ever yet appeared, and we cannot too strongly recommend it to every student of medicine. It is sufficiently comprehensive without being lengthy, and the principles, which are eminently sound, can easily be mastered and understood." — *Medical Record.*

Bouchardat's Annual Abstract

OF THERAPEUTICS, MATERIA MEDICA, PHARMACY, AND TOXICOLOGY, FOR 1867, *with an Original Memoir of Gout, Gravel, Urinary Calculi, &c.* By A. BOUCHARDAT, Professor of Hygiene to the Faculty of Medicine, Paris, &c. Translated and Edited by M. J. DE ROSSET, M.D., Adjunct to the Professor of Chemistry in the University of Maryland. In one Volume. Price, in cloth, . . $1.50.

Andrews' Hand-Book of the Practice of Medicine. *In Preparation.*

Bull on the Maternal Management of Children in Health and Disease.

A New and Improved Edition. By THOMAS BULL, M.D., *Member of the Royal College of Physicians.* 12mo. $1.25

Reports on the Progress of Medicine, Surgery, and the Allied Sciences. Containing:

1. REPORT ON PHYSIOLOGY, by Henry Power, F.R.C.S., M.B., Lond.
2. REPORT ON PRACTICAL MEDICINE, by Francis Edmund Anstie, M.D., etc.
3. REPORT ON SURGERY, by T. Holmes, Esq., M.A., F.R.C.S., etc.
4. REPORT ON OPHTHALMIC MEDICINE AND SURGERY, by Thomas Windsor, M.D.
5. REPORT ON MIDWIFERY AND THE DISEASES OF WOMEN AND CHILDREN, by R. Barnes, M.D., F.R.C.P.
6. REPORT ON MEDICAL JURISPRUDENCE, by C. Hilton Fagge, M.D., F.R.C.P.
7. REPORT ON MATERIA MEDICA AND GENERAL THERAPEUTICS, by C. Hilton Fagge, M.D., F.R.C.P.
8. REPORT ON PUBLIC HEALTH, by C. Hilton Fagge, M.D., etc.

In One Volume, handsomely bound in cloth, Price, . . $2 00

"This volume, published under the auspices of the New Sydenham Society, now makes its welcome appearance, at a low price, in the United States. It is the most carefully prepared retrospect with which we are acquainted. Each department is in charge of a gentleman of reputation, and everything is done to summarize, in a readable way, all the more important advances of medicine over the globe. It is handsomely bound and elegantly printed."—*N. Y. Medical Record.*

Birch on Constipation. From the Third London Edition.

CONSTIPATED BOWELS; the various Causes and the Different Means of Cure. By S. B. BIRCH, M.D., *Member of the Royal College of Physicians of London, &c.* The Third Edition. One volume, 16mo. Price, ... $1.00

Braithwaite's Epitome of the Retrospect of Practical Medicine and Surgery.

Two large Octavo Volumes of 900 pages each, . . . $10.00

Braithwaite's Retrospect of Practical Medicine and Surgery.

"The cream of medical literature."

Published half-yearly, in January and July, at $2.50 per annum, if paid in advance; or $1.50 for single parts.

British and Foreign Medico-Chirurgical Review, and Quarterly Journal of Practical Medicine and Surgery.

Published in London on the 1st of January, April, July, and October, at 6 shillings per number, or 24 shillings per annum, and furnished in this country at $10.00 *per annum; being much less than the present cost of importation of English books. Containing Analytical and Critical Reviews, a Bibliographical Record, Original Communications, and a Chronicle of Medical Science, chiefly Foreign and Contemporary.*

This is considered the leading Medical Review in the English language. It is everywhere looked upon as high authority. It presents in its pages a large amount of valuable and interesting matter, and will post the physician who reads it, fully up to the present day in medical literature.

Chambers's Lectures on the Renewal of Life.

A New American from the Fourth London Edition.

Lectures chiefly clinical, illustrative of a Restorative System of Medicine. By THOS. K. CHAMBERS, M. D., *Physician to St. Mary's Hospital.* Author of "The Indigestions," &c., &c. Octavo, . . $5.00

"The medical profession in this country are under obligations to the American publishers for this reprint of Dr. Chambers' Lectures — a work whose time is forever, everywhere in its place, admirable in tone, full of valuable instructions and practical teachings, and written in clear, compact, and often epigrammatic English. We can offer but a brief notice of this intrinsically good book, which is certain of finding a wide circle of readers, and we should hope a place in every medical library."— *New York Medical Journal.*

"This work is of the highest merit, written in a clear, masterly style, and devoid of technicalities. It is simply what it professes to be, Lectures Clinical, delivered from cases observed at the bedside; therefore more valuable as enunciating the views and experiences of a practical mind aided by actual observation. They are of deep interest, and replete with facts having a practical bearing, and will well repay perusal. . . . . We can recommend Dr. Chambers' book freely and with confidence, as the work of a great mind practical in its bearing, and simple to the understanding of all."— *Canada Medical Journal.*

Chew on Medical Education.

A Course of Lectures on the Proper Method of Studying Medicine. By SAMUEL CHEW, M.D., *Professor of the Practice and Principles of Medicine and of Clinical Medicine in the University of Maryland.* 12mo. $1.00

"The topics discussed in this volume are of books — of time to be devoted to study — and the manner — of the order of medical studies — of the taking of notes — of clinical instruction — dissections — auscultation — medical schools, &c.

"Dr. Chew was an eminent member of the medical profession, and a well-known teacher of medicine. He was, therefore, well fitted for the judicious performance of this task, upon which he seems to have entered with interest and pleasure. It is a well-timed book, and will serve as a most excellent manual for the student, as well as a refreshing and suggestive one to the practitioner." — *Lancet and Observer.*

Cazeaux's Great Work on Obstetrics.
The Fifth American Edition. 175 Illustrations.

A Theoretical and Practical Treatise on Midwifery. Including the Diseases of Pregnancy and Parturition, and the attention required by the Child from its Birth to the Period of Weaning. By P. CAZEAUX, *Member of the Imperial Academy of Medicine, Adjunct Professor in the Faculty of Medicine of Paris, &c., &c.* Revised and annotated by S. TARNIER, *Adjunct Professor to the Faculty of Medicine, Paris, &c., &c.* Translated by W. R. BULLOCK, M. D. With new Lithographic and other Illustrations on Wood. One volume Royal Octavo, of over 1100 pages.

  Price, bound in Cloth, Bevelled Boards, ... $6.50
    " " Leather, ... 7.50

"Written expressly for the use of students of medicine, and those of midwifery especially, its teachings are plain and explicit, presenting a condensed summary of the leading principles established by the masters of the obstetric art, and such clear, practical directions for the management of the pregnant, parturient, and puerperal states, as have been sanctioned by the most authoritative practitioners, and confirmed by the author's own experience. Collecting his materials from the writings of the entire body of antecedent writers, carefully testing their correctness and value by his own daily experience. and rejecting all such as were falsified by the numerous cases brought under his own immediate observation, he has formed out of them a body of doctrine, and a system of practical rules, which he illustrates and enforces in the clearest and most simple manner possible."—*Examiner.*

Canniff's Manual of the Principles of Surgery.

Based on Pathology, for Students, by WM. CANNIFF, *Licentiate of the Medical Board of Upper Canada; M.D. of the University of New York; M.R.C.S. of England; formerly House Surgeon to the Seamen's Hospital, New York; late Professor of General Pathology and the Principles and Practice of Surgery, University Victoria College.* C. W. Octavo. $4.50

"This manual is evidently the production of a man who is well informed on his subject, and who moreover has had experience as a teacher and as a practitioner. He has profited by the study of the best authors on the principles of surgery, tested practically their doctrines, and has presented his own views, well arranged and clearly expressed, for the advantage of others."—*American Journal of Med. Science.*

Cleaveland's Pronouncing Medical Lexicon.

A NEW AND IMPROVED EDITION (THE ELEVENTH).

Containing the Correct Pronunciation and Definition of most of the Terms used by Speakers and Writers of Medicine and the Collateral Sciences. By C. H. CLEAVELAND, M.D., *Member of the American Medical Association, etc., etc.* A small Pocket Volume. $1.25

This little work is both brief and comprehensive; it is not only a Lexicon of all the words in common use in Medicine, but it is also a Pronouncing Dictionary. a feature of great value to Medical Students. To the Dispenser it will prove an excellent aid, and also to the Pharmaceutical Student. This edition contains a List of the Abbreviations used in Prescriptions, together with their meaning; and also of Poisons and their Antidotes: two valuable additions. It has received strong commendation both from the Medical Press and from the profession.

Cohen on Inhalation.

Its Therapeutics and Practice. A Treatise on the Inhalation of Gases, Vapors, Nebulized Fluids, and Powders; including a Description of the Apparatus employed, and a Record of Numerous Experiments, Physiological and Pathological; with Cases and Illustrations. By I. SOLIS COHEN, M.D. 12mo. Price, ... $2 50

"We recognize in this book the work of a persevering Physician who has faithfully studied his subject, and added to its literature much that is useful from his own experience. It treats respectively of the inhalations of nebulized fluids; of medicated airs, gases, and vapors, and of powders. Dr. Cohen has given us briefly and clearly whatever is valuable in relation to the insufflation of powders in respiratory affections, with the experimental proofs and pathological evidence of their penetration into the bronchial tubes and lung tissues." — *American Journal of Medical Science,* July, 1868.

Prof. Carson's University of Pennsylvania.

A History of the Medical Department of the University of Pennsylvania, from its Foundation in 1765: With Sketches of the Lives of Deceased Professors. By JOSEPH CARSON, M.D., Professor of Materia Medica and Pharmacy in the University of Pennsylvania; Member of the American Philosophical Society, etc. In one volume octavo. Cloth. Price, ... $2 00

"The history of the University of Pennsylvania has a national as well as a local interest, from the early date of its origination, and the connection with it of men of illustrious public reputation, such as Drs. Franklin, Rush, Physick, Gibson, Dewees, Chapman, Wood, &c., &c.

"For fidelity and carefulness of statement and maintenance of the dignity of the Institution, as well as for skill in literary execution, the task of extending and continuing this record could have been confided to no better hands than those of Professor Carson.

"For the labor and love which he has spent in preparing this most interesting and valuable work, Prof. Carson has earned the gratitude of the alumni of the University, and of all others interested in medical education in this country." — *American Journal of Medical Science.*

Dixon on the Eye. } A New Edition, thoroughly Revised, and a great portion Re-written.

A Guide to the Practical Study of Diseases of the Eye, with an Outline of their Medical and Operative Treatment, with Test Types and Illustrations. By JAMES DIXON, F.R.C.S., Surgeon to the Royal London Ophthalmic Hospital, &c., &c. In one volume. Price, . $2 50

"Mr. Dixon's book is essentially a practical one, written by an observant author, who brings to his special subject a sound knowledge of general Medicine and Surgery." — *Dublin Quarterly.*

"Our object is not to review, but to recommend this work to students, with the confident assurance that they will rarely be disappointed in their appeals to it as a reliable guide to the practical study of the Diseases of the Eye." — *American Medical Journal.*

"We have taken great pleasure in a careful perusal of this book, which, both in style and matter, is unsurpassed in any language. It embraces quite a wide range of topics, and furnishes a very valuable practical guide in the medical and surgical treatment of diseases of the eye." — *Buffalo Medical Journal.*

Durkee on Gonorrhœa and Syphilis.

The Fifth Edition, Revised and Enlarged, with Portraits and Colored Illustrations.

A Treatise on Gonorrhœa and Syphilis. By SILAS DURKEE, M.D., *Fellow of the Massachusetts Medical Society, &c., &c. A New and Revised Edition, with Eight Colored Illustrations.* Octavo. . . $5.00

This work of Dr. DURKEE's has received the unqualified approval of the Medical Press and the Profession both in this country and in Europe. The author has devoted himself especially to the treatment of this class of diseases, and his 25 or 30 years experience in doing so is here recorded. No one reading his work can fail in receiving very valuable information from it.

"It is the work of a practical man, the subject is treated in a plain, shrewd manner. The book is a good one, and the therapeutics are laid down with discrimination." — *London Medical Times and Gazette.*

"Dr. Durkee's production is one of those, the perusal of which impresses the reader in favor of the author. The general tone, the thorough honesty everywhere evinced, the philanthropic spirit observable in many passages, and the energetic advocacy of professional rectitude, speak highly of the moral excellence of the writer; nor is the reader less attracted by the skill with which the book is arranged, the manner in which the facts are cited, the clever way in which the author's experience is brought in, and the lucidity of the reasoning, the frequent and extremely fair allusions to the labors of others, and the care with which the therapeutics of venereal complaints are treated." — *Lancet.*

Fuller on Rheumatism, Rheumatic Gout, and Sciatica. A NEW EDITION PREPARING.

Their Pathology, Symptoms, and Treatment. By HENRY WILLIAM FULLER, M.D., *Fellow of the Royal College of Physicians, London; Physician to St. George's Hospital, etc. From the last London Edition.* Octavo. . . . .

Graves' Clinical Lectures on the Practice of Medicine. By ROBERT JAMES GRAVES, M.D., F.R.S., *Professor of the Institutes of Medicine in the School of Physic in Ireland.* Edited by J. MOORE MELIGAN, M.D. *From the Second Revised and Enlarged Edition.* Complete in One Volume. Octavo. Price, . . $6.00

Goff's Combined Day-Book, Ledger, and *Daily Register of Patients, combining not only the Accuracy and Essential Points of a regular Day-Book and Ledger System, without any of the labor and responsibility, but is also a Daily Register of Patients, &c., &c. A large Quarto Volume, strongly bound in half-russia.* Price, $12.00

The advantages of this book are — The account of a *whole* family for *an entire year* can be kept in a very small space. (See Mitchell's account.) No transfer of accounts from one book to another, or from one part of the book to another. No protracted search for an account when wanted. Shows the exact state of an account at any moment.

Gross' American Medical Biography of the Nineteenth Century.

Edited by SAMUEL D. GROSS, M.D., *Professor of Surgery in the Jefferson Medical College, Philadelphia, &c., &c.* With a Portrait of BENJAMIN RUSH, M.D. Octavo. $3.50

Greenhow on Bronchitis, *especially as Connected with Gout, Emphysema, and Diseases of the Heart.* By E. HEADLAM GREENHOW, M.D., *Fellow of the Royal College of Physicians, &c., &c.*

Price, $2.00

"In vivid pictures of the sort of cases which a practitioner encounters in his daily walks, and in examples of the way in which a student ought to turn them over in his mind and make them tools for self-improvement, we have rarely seen a volume richer." — *Brit. and For. Medico-Chirurg. Review.*

Garratt's (Alfred C.) Guide for Using Medical Batteries.

Showing the most approved Apparatus, Methods, and Rules for the Medical Employment of Electricity in the Treatment of Nervous Diseases, &c., &c. With numerous Illustrations. One Volume, octavo. . . $2.00

"The large work on the same subject, and by the same author, is pretty well known to the Profession, but it is bulky and cumbrous, and by no means so practically useful. The present comparatively brief volume contains every thing of importance in regard to the various apparatuses useful to the Medical Electrician and the various modes of application for therapeutic purposes." — *Lancet and Observer.*

Hewitt on the Diseases of Women.
SECOND EDITION, REWRITTEN AND ENLARGED.

The Diagnosis and Treatment of Diseases of Women, including the Diagnosis of Pregnancy. Founded on a Course of Lectures delivered at St. Mary's Hospital Medical School. By GRAILY HEWITT, M. D. Lond., M. R. C. P., *Physician to the British Lying-in Hospital; Lecturer on Midwifery and Diseases of Women and Children at St. Mary's Hospital Medical School; Honorary Secretary to the Obstetrical Society of London, &c.* With a new Series of Illustrations.

Price, in cloth, $5.00; in leather, $6.00.

Hillier's Clinical Treatise on the Diseases of Children. By THOMAS HILLIER, M.D., *Physician to the Hospital for Sick Children, and to University College Hospital, &c., &c.* Octavo.

Price, $3.00

"Our space is exhausted, but we have said enough to indicate and illustrate the excellence of Dr. Hillier's volume. It is eminently the kind of book needed by all medical men who wish to cultivate clinical accuracy and sound practice." — *London Lancet.*

Headland on the Action of Medicines in the System.

By F. W. HEADLAND, M.D., *Fellow of the Royal College of Physicians, &c., &c. Sixth American from the Fourth London Edition. Revised and enlarged.* One Volume, octavo. $3.00

Dr. Headland's work has been out of print in this country nearly two years, awaiting the revision of the author, which now appear in this edition. It gives the only scientific and satisfactory view of the action of medicine; and this not in the way of idle speculation, but by demonstration and experiments, and inferences almost as indisputable as demonstrations. It is truly a great scientific work in a small compass, and deserves to be the handbook of every lover of the Profession. It has received the most unqualified approbation of the *Medical Press*, both in this country and in Europe, and is pronounced by them to be the most *original* and practically useful work that has been published for many years.

Hille's Pocket Anatomist.

Being a Complete Description of the Anatomy of the Human Body; for the Use of Students. By M. W. HILLES, *formerly Lecturer on Anatomy and Physiology at the Westminster Hospital School of Medicine.*
Price, in cloth, ... $1.00
" in Pocket-book form, ... 1.25

Heath on the Injuries and Diseases of the Jaws.

The Jacksonian Prize Essay of the Royal College of Surgeons of England, 1867. By CHRISTOPHER HEATH, F. R. C. S., *Assistant Surgeon to University College Hospital, and Teacher of Operative Surgery in University College. Containing over 150 Illustrations.* Octavo. Price, $6.00

Hodge on Fœticide, or Criminal Abortion.

By HUGH L. HODGE, M.D., *Emeritus Professor in the University of Pennsylvania. A Small Pocket Volume.* Price in paper covers, 30
" flexible cloth, 50

This little book is intended to place in the hands of professional men and others the means of answering satisfactorily and intelligently any inquiries that may be made of them in connection with this important subject.

Holmes' Surgical Diseases of Infancy and Childhood.

By J. HOLMES, M.A., *Surgeon to the Hospital for Sick Children, &c. Second Edition. Revised and Enlarged.* Octavo. Price, $9.00

Hufeland's Art of Prolonging Life.

Edited by ERASMUS WILSON, M.D., F R.S. *Author of "A System of Human Anatomy," "Diseases of the Skin," &c., &c.* 12mo. Cloth. $1.25

Mackenzie on the Laryngoscope, Diseases of the Throat, &c. Second Edition.

The Use of the Laryngoscope in Diseases of the Throat. With additions, and an Essay on Hoarseness, Loss of Voice, and Stridulous Breathing in relation to Nervo-Muscular affections of the Larynx, by MORELL MACKENZIE, M.D., *Physician to the Hospital for Diseases of the Throat, &c., &c. Second Edition, with additions, and a Chapter on the Nasal Passages,* by J. SOLIS COHEN, M.D., *Author of "Inhalation, Its Therapeutics and Practice," &c. Illustrated by two lithographic plates, and 51 engravings on wood.* Octavo. Price, ... $3.00

"While laryngoscopy was in its infancy, and before it had begun to engage to any extent the attention of the Profession, it was studied with the greatest care and enthusiasm by the author of this treatise. A personal friend of Czermak's, who has done more than any other continental physician to introduce the laryngoscope into practice, he has profited by the opportunities which he thus possessed of becoming acquainted with the anatomy and morbid anatomy of the larynx. But he has done much more than this. As will be seen by a perusal of this treatise, he has modified the instruments at present in use for the examination of the larynx, and has invented others for therapeutical purposes. Those who are anxious to study the diseases of the larynx and the mode of using the laryngoscope, cannot do better than purchase the treatise before us, as it is by far the best which has been published, and is thoroughly to be relied upon."—*Glasgow Medical Journal.*

Morris on the Pathology and Therapeutics of Scarlet Fever.

By CASPER MORRIS, M.D., *Fellow of the College of Physicians of Philadelphia, &c., &c.* A New Enlarged Edition. Octavo. . $1.50

Meigs and Pepper's Practical Treatise on the Diseases of Children.

Fourth Edition, thoroughly Revised and greatly Enlarged.
By J. FORSYTH MEIGS, M. D., *Fellow of the College of Physicians of Philadelphia, &c., &c., and* WILLIAM PEPPER, M D., *Physician to the Philadelphia Hospital, &c., &c.,* forming a Royal Octavo Volume of over 900 pages. Price, bound in Cloth, . . . $6.00
" " Leather, . . . 7.00

Dr. Meigs' work has been out of print for some years. The rapid sale of the three previous editions, and the great demand for a new edition, is sufficient evidence of its great popularity; while the very large practice of many years' standing of the author in the speciality of "Diseases of Children," imparts to it a value unequalled, probably, by any other work on the same subject now before the Profession. This present edition has been almost entirely rewritten and rearranged, and no effort or labor has been spared by either Drs. Meigs or Pepper, to make it represent fully in its most advanced state the present condition of Medicine as applied to Children's Diseases.

Murphy's Review of Chemistry for Students.

Adapted to the Courses as Taught in the Principal Medical Schools in the United States. By JOHN G. MURPHY, M.D. *In One Volume.* $1.25.

"This is an exceedingly well-arranged and convenient Manual. It gives the most important facts and principles of Chemistry in a clear and very concise manner, so as to subserve most admirably the object for which it was designed."—*North Western Medical and Surgical Journal.*

Maxson's Practice of Medicine.

A New Text-Book on the Practice of Medicine. By EDWIN R. MAXSON, M.D., *formerly Lecturer on the Institutes and Practice of Medicine in the Geneva Medical College. In One Volume.* Royal 8vo. . $4.00

"Judging from his work, he must be a correct observer, of plain, strong common sense, having the progress and perfection of the healing art, and the amelioration of suffering, earnestly at heart, free from prejudice, and open to conviction. The fact of employing, and thereupon recommending valuable remedial agents, as yet, for various reasons, under the ban, and misunderstood by many physicians, is an honor to him, and gives a certain additional value to his book." — *American Medical Monthly.*

Mendenhall's Medical Student's Vade Mecum.

A Compendium of Anatomy, Physiology, Chemistry, The Practice of Medicine, Surgery, Obstetrics, Diseases of the Skin, Materia Medica, Pharmacy, Poisons, &c., &c. By GEORGE MENDENHALL, M.D., *Professor of Obstetrics in the Medical College of Ohio, Member of the American Medical Association, &c., &c. The Eighth Edition, Revised and Enlarged; with 224 Illustrations.* $2.50

"This volume puts the student in possession of a condensed medical library. Its accuracy is a strong recommendation, while the portability of a volume containing the whole circle of medical science is a matter that will have weight with those for whose service the book was originally designed. The work is offered, too, extremely cheap, and will be found a valuable assistant even to a well-informed practitioner of any branch of medicine." — *Boston Medical and Surgical Journal.*

Paget's Lectures on Surgical Pathology.

Delivered at the Royal College of Surgeons of England, by JAMES PAGET, F.R.S., *Surgeon to Bartholomew and Christ's Hospital, &c., &c. The Third American from the Second London Edition, Edited and Revised* by WILLIAM TURNER, M.B., *Lond. Senior Demonstrator of Anatomy in the University of Edinburgh, &c., &c. In One Volume, Royal Octavo; with Numerous Illustrations.*

Price, in bevelled cloth, ... $6.00
" in leather, ... 7.00

Pennsylvania Hospital Reports. *Edited by a Committee of the Hospital Staff,* J. M. DACOSTA, M.D., *and* WILLIAM HUNT, M.D. *Vols. 1 and 2, for 1868 and 1869, each volume containing upwards of* Twenty *Original Articles, by former and present Members of the Staff, now eminent in the Profession, with Lithographic and other Illustrations.*

Price per volume, ... $4.00

At last, however, the work has been commenced, the Philadelphia Physicians being the first to occupy this field of usefulness, having issued the first volume of the Reports of the above hospital last year, and the second volume on January 1st, 1869. The first Reports were so favorably received on both sides of the Atlantic, that it is hardly necessary to speak for this volume the universal welcome of which it is deserving. We cannot close our remarks without stating that the papers are all valuable contributions to the literature of medicine, reflecting great credit upon their authors, and the work is one of which the Pennsylvania Hospital may well be proud. It will do much toward elevating the profession of this country in the estimation of their foreign brethren."
—*American Journal of Obstetrics,* May, 1869.

Pereira's Physician's Prescription Book.

Containing Lists of Terms, Phrases, Contractions, and Abbreviations, used in Prescriptions, with Explanatory Notes, the Grammatical Constructions of Prescriptions, Rules for the Pronunciation of Pharmaceutical Terms, A Prosodiacal Vocabulary of the Names of Drugs, etc., and a series of Abbreviated Prescriptions illustrating the use of the preceding terms, etc.; to which is added a Key, containing the Prescriptions in an unabbreviated Form, with a Literal Translation, intended for the use of Medical and Pharmaceutical Students. By JONATHAN PEREIRA, M.D., F.R.S., etc. From the Fourteenth London Edition.

Price, in cloth, ... $1.25
" in leather, with Tucks and Pocket, . . . 1.50

This little work has passed through fourteen editions in London and several in this country. The present edition of which this is a reprint has been carefully revised and many additions made to it. Its great value is proven both by its large sale and the many favorable notices of it in the Medical Press.

Physicians Visiting List. Published annually.

SIZES AND PRICE.

For 25 Patients weekly.	Tucks, pockets, and pencil,					$1 00
50 "	" "	"				1 25
75 "	" "	"				1 50
100 "	" "	"				2 00
50 "	" 2 vols. {Jan. to June. / July to Dec.}	"				2 50
100 "	" 2 vols. {Jan. to June. / July to Dec.}	"				3 00

Also, AN INTERLEAVED EDITION,

for the use of *Country Physicians* and others who compound their own Prescriptions, or furnish *Medicines* to their patients. The additional pages can also be used for *Special Memoranda*, recording important cases, &c., &c.

For 25 Patients weekly, interleaved, tucks, pockets, etc., ... $1 50
50 " " " " " 1 75
50 " " 2 vols. {Jan. to June. / July to Dec.} " " 3 00

Prince's Orthopedic Surgery.

ORTHOPEDICS: *A Systematic Work upon the Prevention and Cure of Deformities.* By DAVID PRINCE, M.D. *With Numerous Illustrations.* Octavo. $3.00

"This is a good book, upon an important practical subject; carefully written, abundantly illustrated, and well printed. It goes over the whole ground of deformities of all degrees — from cleft-palate and club-foot, to spinal curvatures and ununited fractures. It appears, moreover, to be an original book, so far as one chiefly of compilation can be so. Such a book was wanted, and it deserves success." — *Med. & Surg. Reporter.*

Prince's Plastic Surgery.

A New Classification and a Brief Exposition of Plastic Surgery. By DAVID PRINCE, M. D. *In One Volume Octavo. With Numerous Illustrations.* Price, ... $1.50

Radcliffe's Lectures on Epilepsy, Pain, Paralysis,

And certain other Disorders of the Nervous System, by CHARLES BLAND RADCLIFFE, M.D., *Fellow of the Royal College of Physicians of London, Physician to the Westminster Hospital,* etc., etc. With Illustrations. 12mo. $2.00

"The reputation which Dr. Radcliffe possesses as a very able authority on nervous affections, will commend his work to every medical practitioner. Disorders of the nervous system are very imperfectly comprehended, much concerning them being involved in mystery; and while Dr. Radcliffe has taken advantage of the ample room to theorize, which his subject afforded, he has not failed to bring forward strong and formidable facts to prove the deductions he attempts to draw. We recommend it to the notice of our readers as a work that will throw much light upon the Physiology and Pathology of the Nervous System."—*Canada Medical Journal.*

Robertson's Manual on Extracting Teeth.

Founded on the Anatomy of the Parts involved in the Operation; the Kinds and Proper Construction of the Instruments to be used; the Accidents liable to occur from the Operation, and the Proper Remedies to retrieve such Accidents. By ABRAHAM ROBERTSON, D.D.S., M.D., *Author of "Prize Essay on Extracting Teeth,"* &c. In One Volume, with Illustrations. Second Edition. Revised and Improved. . . $1.50

"The author is well known as a contributor to the literature of the Profession; and, as a clear, terse, forcible writer, he has devoted considerable care to the subject, and treated it with his usual ability. The work is valuable, not only to the dental student and practitioner, but also to the medical student and surgeon; and especially so to the military surgeon, who, in times like the present, is called upon so frequently to perform the operation of extracting teeth."—*Dental Cosmos.*

Ranking's Half-yearly Abstract of the Medical Sciences.

Price, per annum, if paid in advance, $2.50. Half-yearly volumes, $1.50. The first thirty-two volumes, bound in sixteen volumes, leather, can be furnished each at $2.00. Half-yearly volumes, in paper covers, from 1 to 34, each at $1.00.

Renouard's History of Medicine.

History of Medicine from its origin to the Nineteenth Century. With an Appendix containing a Philosophical and Historical Review of Medicine to the present time. By P. V. RENOUARD, M.D. *Translated from the French by* CORNELIUS G. COMEGYS, M.D., *Professor of the Institutes of Medicine in the Medical College of Ohio,* etc. In One Volume Octavo. Price, ... $4.00

"From the pages of Dr. Renouard, a very accurate acquaintance may be obtained with the history of medicine—its relation to civilization, its progress compared with other sciences and arts, its most distinguished cultivators with the several theories and systems proposed by them, and its relationship to the reigning philosophical dogmas of the several periods. His historical narration is clear and concise, tracing the progress of medicine through its three ages or epochs—that of foundation or origin, that of tradition, and that of renovation."—*American Journal of Medical Science.*

"The best history of medicine extant, and one that will find a place in the library of every physician who aims at an acquaintance with the past history of his profession. There are many items in it we should like to offer for the instruction and amusement of our readers."—*American Journal of Pharmacy.*

Ryan's Philosophy of Marriage.

In its Social, Moral, and Physical Relations, with an Account of the Diseases of the Genito-Urinary Organs. The Physiology of Generation in the Animal and Vegetable Kingdoms, &c., &c. By MICHAEL RYAN, M.D., Member of the Royal College of Physicians and Surgeons in London, &c. 12mo. $1.00

"Dr. Ryan is above reproach or suspicion; and with a singular degree of candor and independence, explains, in a true and philosophical manner, every branch of the subject which he considers essential to be understood by all intelligent persons." — *Boston Medical and Surgical Journal.*

Reese's Analysis of Physiology.

Being a Condensed View of the most important Facts and Doctrines, designed especially for the Use of Students. By JOHN J. REESE, M.D., Professor of Medical Jurisprudence, including Toxicology, in the University of Pennsylvania, &c., &c. Second Edition, Enlarged. 12mo. $1.50.

Reese's American Medical Formulary.
12mo. $1.50

Reese's Syllabus of Medical Chemistry.
$1.00

Stillé's Epidemic Meningitis;

Or, Cerebro-Spinal Meningitis. By ALFRED STILLÉ, M.D., Professor of the Theory and Practice of Medicine in the University of Pennsylvania. &c. &c. In One Volume Octavo. $2.00

"This monograph is a timely publication, comprehensive in its scope, and presenting within a small compass a fair digest of our existing knowledge of the disease, particularly acceptable at the present time. It is just such a one as is needed, and may be taken as a model for similar works." — *Am. Journal Med. Sciences.*

Sydenham Society's Publications. New Series, 1859

to 1870 inclusive, 12 years, 50 vols. Subscriptions received, and back years furnished at $10.00 per year. Full prospectus, with the Reports of the Society, and Lists of Books published, furnished free upon application.

Stillé's Elements of General Pathology.

A Practical Treatise on the Causes, Forms, Symptoms, and Results of Disease. By ALFRED STILLÉ, M.D., Professor of the Theory and Practice of Medicine in the University of Pennsylvania, &c. (In Preparation.)

Sansom on Chloroform.

Its Action and Administration, by ARTHUR ERNEST SANSOM, M.B., Physician to King's College Hospital, etc., etc. 12mo. . . $2.00

"The work of Dr. Sansom may be characterized as most excellent. Written not alone from a theoretical point of view, but showing very considerable experimental study, and an intimate clinical acquaintance with the administration of these remedies,— passing concisely over the whole ground, giving the latest information upon every point,— it is just the work for the student and practitioner. The author may rest assured that, although in his preface he objects to the 'hackneyed expression of endeavoring to supply a want,' this is just what he has done — supplied and well supplied a want, for no such book existed before in our language."— *American Medical Journal.*

Scanzoni's Practical Treatise on the Diseases of the Sexual Organs of Women.

Translated from the French of Drs. H. DOR and A. SOCIN, *and annotated with the approval of the authors.* By A. K. GARDNER, A.M., M.D., *Professor of Clinical Midwifery, &c., &c., in the New York Medical College.* With Numerous Illustrations. Octavo. . . . $5.00

In the etiology, pathology, and therapeutics of female diseases, with all the improvements which have been realized during the last twenty years, this volume is exceedingly rich; while in its arrangement it is so methodical that it must constitute one of the best text-books for students, and one of the most reliable aids to the busy practitioner.

Stokes on the Diseases of the Heart and the Aorta.

By WILLIAM STOKES, *Regius Professor of Physic in the University of Dublin; Author of the Treatment and Diagnosis of the Diseases of the Chest, &c., &c.* Second American Edition. Octavo. . . $3.00

Spratt's Obstetrical Tables.

Comprising Graphic Illustrations, with Descriptions and Practical Remarks exhibiting on Dissected Plates many important subjects in Midwifery. By G. SPRATT, *Surgeon Accoucheur.* First American from the Fourth and Greatly Improved London Edition, carefully Revised, with Additional Notes and COLORED Plates. One Volume Quarto. Price, ... $8.00

Skoda on Auscultation and Percussion.

By JOSEPH SKODA. *Translated from the Fourth German Edition, by* W. O. MARKHAM, M.D., *Assistant Physician to St. Mary's Hospital.* 12mo. $1.50

Tanner's Practice of Medicine.

FIFTH AMERICAN EDITION.

The *Practice of Medicine*, by THOMAS HAWKES TANNER, M.D., *Fellow of the Royal College of Physicians, Author of A Practical Treatise on the Diseases of Infancy and Childhood, etc., etc. Fifth American from the Sixth London Edition. Greatly Enlarged and Improved.*

Price, bound in cloth, ... $6.00
" " in leather, ... 7.00

Dr. Tanner's work on the Practice of Medicine is so well known in this country, and has had such an extensive and rapid sale, that it seems almost unnecessary to say anything in reference to it; the present edition, however, contains such substantial additions and alterations as almost to constitute it a new work, and from being a comparatively small volume it now forms a handsome octavo of nearly 1000 pages; all that was useful and practical in the smaller volume has been retained and much new matter added, written in the same condensed and easy style.

"The leading feature of this book is its essentially practical character. Dr. Tanner has produced a more complete System of Medicine than any with which we are acquainted. It is the result of long experience and hard practice, and it is therefore valuable as a guide, and trustworthy as an exemplar." — *London Lancet.*

Tanner's Practical Treatise on the Diseases of Infancy and Childhood.

Octavo. $3.00

This book differs from other works of the kind, in embracing a wider range of subjects than is usually contained in treatises on children's diseases; besides the ordinary complaints of those subjects, it includes many affections which, though common to adults and children, yet offer some modification in form, or in the indications for treatment, when occurring in the latter. Thus, we have an account of diseases of the eye, ear, and skin, of small-pox, scrofula, tuberculosis, syphilis, bronchocele, and cretinism, diseases of the kidneys and genital organs, and some of the accidents common to childhood. The style of the work is condensed, and the book might with truth be called a manual, rather than a treatise, but there is nothing superficial about it; — everything really important is given, while the discussion of disputed subjects, and, in fact, of everything which is not of practical importance in the study and treatment of children's diseases, is omitted.

Tanner's Index of Diseases and their Treatment.

With *upwards of 500 Formulæ for Medicines, Baths, Mineral Waters, Climates for Invalids, &c., &c.* Octavo. $3.00

"Dr. Tanner has been peculiarly happy in appreciating and supplying the wants of the Profession. His Index of Diseases gives the derivation of words after the manner of a good Medical Dictionary; an outline of every disease, including many surgical diseases, with their symptoms and mode of treatment; an admirable collection of Formulæ, and an account of the climates of the various parts of the world suitable for invalids. It also contains at the beginning of the work a tabular synopsis of subjects, which does double duty at once, a Nosology and an index. It will be found a most valuable companion to the judicious practitioner." — *Lancet.*

Tanner's Memoranda of Poisons.

From the Second London Edition. $0.50

Trousseau's Lectures on Clinical Medicine.

Delivered at the Hotel Dieu, Paris, by A. TROUSSEAU, Professor of Clinical Medicine in the Faculty of Medicine, Paris. Translated and edited, with Notes and Appendices, by P. VICTORE BAZIRE, M.D., Assistant Physician to the National Hospital for the Paralyzed and Epileptic, &c.
Volume One. Cloth, ... 5 00
Volume Two, ... 5 00
Volume Three, now Ready, ... 5 00

"This book furnishes us with an example of the best kind of clinical teaching, and we are much indebted to the translator for supplying the Profession with these admirable Lectures. It is a book which deserves to be popularized. We scarcely know of any work better fitted for presentation to a young man when entering upon the practical work of his life. The delineation of the recorded cases is graphic, and their narration devoid of that prolixity which, desirable as it is for purposes of extended analysis, is highly undesirable when the object is to point to a practical lesson." — *London Medical Times and Gazette.*

Tyler Smith's Obstetrics.

A Course of Lectures. By WILLIAM TYLER SMITH, M.D., Physician, Accoucheur, and Lecturer on Midwifery, and the Diseases of Females, in St. Mary's Hospital, Medical School, &c., &c. With Numerous Illustrations. Edited by A. K. GARDNER, M.D., Fellow of the New York Academy of Medicine, &c., &c. Octavo. $5.00

Toynbee on Diseases of the Ear.

Their Nature, Diagnosis, and Treatment. A new London Edition, with a Supplement. By JAMES HINTON, Aural Surgeon to Guy's Hospital, &c. With Illustrations. Octavo. . . . . Price, $5.00

Thompson's Clinical Lectures on Pulmonary Consumption.
Octavo. $2.00.

Tyson's Cell Doctrine:

Its History and Present State, with a Copious Bibliography of the Subject, for the use of Students of Medicine and Dentistry. By JAMES TYSON, M.D., Lecturer on Microscopy in the University of Pennsylvania, &c., &c. In One Volume, with a Colored Plate, and numerous Illustrations on Wood. Price, ... $2.00

Tilt's Elements of Health, and Principles of Female Hygiene.

By F. J. TILT, M.D., Senior Physician to the Lying-in Charity, Author of Works on the Diseases of Menstruation, Uterine Therapeutics, &c., &c. 12mo. $1 50

"Dr. Tilt divides life into the septennial epochs so long adopted by philosophers and medical men, discussing, under the different ages, the physical and moral relations, diseases, &c., peculiar to each. The chapter devoted to the age from fourteen to twenty-one years contains much valuable advice respecting the menstrual function during that period. Tables showing the value of life at each of the different periods of life, are appended in their proper places; and the work also contains other statistics of value and interest. The whole work has been prepared with great care, and contains a large amount of valuable information, which professional men may consult with profit." — *N. Y. Medical Times.*

Taylor's Theory and Practice of the Movement-Cure.

Or, the Treatment of Lateral Curvature of the Spine, Paralysis, Indigestion, Constipation, Consumption, Angular Curvatures, and other Deformities, Diseases Incident to Women, Derangements of the Nervous System, and other Chronic Affections, by the Swedish System of Localized Movements. By CHARLES TAYLOR, M.D. With Illustrations. 12mo. . $1.50

The work of Dr. Taylor is a systematic treatise, containing the principles on which this treatment is based, and full and explicit directions in their application to individual diseases. The author discusses the nutritive processes, muscular contraction, and the physiology of general exercise, the subjects of the first three chapters, in a most satisfactory manner. The work is purely of a scientific character, and commends itself as such to the attention of all physicians.

Virchow's Cellular Pathology.

As based upon Physiological and Pathological History. Translated from the Second Edition of the Original. By FRANK CHANCE, B.A., M.A., Cantab Licentiate of the Royal College of Physicians, &c., &c. With Notes and Numerous Emendations, principally from MSS. Notes of the Author, and Illustrated by 144 Engravings. Octavo. . . $5.00

Prof. Virchow and his writings are well known wherever the science of medicine is studied. This work has been selected by the Medical Bureau of the United States for general distribution in the hospitals and medical stations of the army; recording, as it does, the researches in this branch of science down to the present time.

The importance of the subject, the new ideas advanced, and the established reputation of the author, induced the publication of this book, and has made it a standard work throughout Europe and in this country.

Virchow on Morbid Tumors.

IN PREPARATION.

Walker on Intermarriage.

Or, the Mode in which, and the Causes why, Beauty, Health, and Intellect result from certain Unions, and Deformity, Disease, and Insanity from others. With Illustrations. By ALEXANDER WALKER, Author of "Woman," "Beauty," &c., &c. 12mo. $1.50

"The author is evidently a careful observer, and a proper thinker, and has presented us with a vast amount of information, derived both from man and the inferior animals. He has aimed to be useful, by pointing out how bodily deformities and mental infirmities may be forestalled; and how marriages among blood relations tend to the degeneracy of the offspring. He also shows how, by carefully assorted marriages, the means of improving general organization and beauty of countenance, as well as mental and physical vigor, are, in a great degree, under the control of man. Although not strictly a medical work, we cannot refrain from commending it to the perusal of the Profession, as it contains much that is valuable in a hygienic point of view."— *Medical Examiner*

Wythes' Physician's Pocket, Dose, and Symptom Book.

Containing the Doses and Uses of all the Principal Articles of the Materia Medica, and Original Preparations; A Table of Weights and Measures, Rules to Proportion the Doses of Medicines, Common Abbreviations used in Writing Prescriptions, Table of Poisons and Antidotes, Classification of the Materia Medica, Dietetic Preparations, Table of Symptomatology, Outlines of General Pathology and Therapeutics, &c. By JOSEPH H. WYTHES, A.M., M.D., &c. *The Eighth Revised Edition.*
Price, in cloth, ... $1.00
" leather, tucks, with pockets, ... 1.25

This little manual has been received with much favor, and a large number of copies sold. It was compiled for the assistance of students, and to furnish a vade mecum for the general practitioner, which would save the trouble of reference to larger and more elaborate works. The present edition has undergone a careful revision. The therapeutical arrangement of the Materia Medica has been added to it, together with such other improvements as it was thought might prove of value to the work.

Waring's Manual of Practical Therapeutics.

Considered chiefly with reference to Articles of the Materia Medica. By EDWARD JOHN WARING, F.R.C.S., F.L.S., &c., &c. *From the Second London Edition.* Royal Octavo.
Price, in cloth, ... $6.00
" in leather, ... 7.00

There are many features in Dr. Waring's Therapeutics which render it especially valuable to the Practitioner and Student of Medicine, much important and reliable information being found in it not contained in similar works; it also differs from them in its completeness, the convenience of its arrangement, and the greater prominence given to the medicinal application of the various articles of the Materia Medica in the treatment of morbid conditions of the Human Body, &c., &c. It is divided into two parts, the alphabetical arrangement being adopted throughout the volume. For the further convenience of the reader there is also added an INDEX OF DISEASES, with a list of the medicines applicable as remedies, and a full INDEX of the medicines and preparations noticed in the work.

"Our admiration, not only for the immense industry of the author, but also of the great practical value of the volume, increases with every reading or consultation of it. We wish a copy could be put in the hands of every student or practitioner in the country. In our estimation it is the best book of the kind ever written." — *N. Y. Medical Journal.*

Weber's Clinical Hand-Book of Auscultation and Percussion.

An Exposition from First Principles of the Method of Investigating Diseases of the Respiratory and Circulating Organs. Translated by JOHN COCKLE, M. D. With Illustrations. Price, . . $1.00

Walton's Operative Ophthalmic Surgery.

By HAYNES WALTON, F.R.C.S., *Surgeon to the Central London Ophthalmic Hospital, &c. With 169 Illustrations. Edited by* S. LITTELL, M.D., *Surgeon to the Wills Hospital for the Diseases of the Eye, &c.* Octavo. $4.00

"It is eminently a practical work, evincing in its author great research, a thorough knowledge of his subject, and an accurate and most observing mind." — *Dublin Quarterly Journal.*

Watson's Practice abridged.

A Synopsis of the Lectures on the Principles and Practice of Physic. Delivered at King's College, London, by THOMAS WATSON, M.D., *Fellow of the Royal College of Physicians, &c., &c. From the last London Edition. With a concise but Complete Account of the Properties, Uses, Preparations, Doses, &c., of all the Medicines mentioned in these Lectures, and other Valuable Additions, by* J. J. MEYLOR, A.M., M.D., &c., &c. *A neat Pocket Volume bound in cloth flexible.* . . . $2.00

Wells' Treatise on the Diseases of the Eye,

illustrated by Ophthalmoscopic Plates and Numerous Engravings on Wood. By J. SŒLBERG WELLS, *Ophthalmic Surgeon to King's College Hospital, &c.* Second London Edition, cloth, $6.50; leather, $7.50.

This is the author's own edition, printed in London under his supervision, and issued in this country by special arrangement with him.

Wright on Headaches.

Their Causes and their Cure. By HENRY G. WRIGHT, M.D., *Member of the Royal College of Physicians, &c. &c.* From the Fourth London Edition. 12mo. Cloth. $1.25

"Few affections are more unmanageable and more troublesome than those of which this essay treats; and we doubt not that any suggestions by which we can relieve them will be gladly received by physicians. The author's plan is simple and practical. He treats of headaches in childhood and youth, in adult life and old age, giving in each their varieties and symptoms, and their causes and treatment. It is a most satisfactory monograph, as the mere fact that this is a reprint of the *fourth* edition, testifies.

"The great pains which the author takes to clear up the differential diagnosis of the different varieties, and establish a satisfactory basis for rational treatment, are everywhere visible. While such a valuable fund of information is offered to the practitioner at the cost of a single visit, he should not let his patient suffer for want of it." — *Medical and Surgical Reporter.*

Wells on Long, Short, and Weak Sight, *and their Treatment by the Scientific Use of Spectacles.* Third Edition Revised, with Additions and Numerous Illustrations. By J. SŒLBERG WELLS. Octavo. . . . . . . Price, $3.00

Harris's Dictionary of Medical Terminology,

DENTAL SURGERY, AND THE COLLATERAL SCIENCES. By CHAPIN A. HARRIS, M.D., D.D.S., *Professor of the Principles of Dental Surgery in the Baltimore College, Member of the American Medical Association, &c., &c. The Third Edition, carefully revised and enlarged, by* FERDINAND J. S. GORGAS, M.D., D.D.S., *Professor of Dental Surgery in the Baltimore College, &c., &c.* Royal octavo. Cloth, $6.50. Leather, $7.50

This Dictionary has been for a long time out of print; a new edition has been much needed by the Profession, a constant and increasing demand existing for it. The present edition has been thoroughly revised by Professor Gorgas, Dr. Harris's successor in the Baltimore Dental College. Many additions and corrections have been made, and some two to three thousand new words added. The doses of the more prominent medicinal agents have also been added, and in every way the book has been greatly improved, and its value enhanced.

Harris's Principles and Practice of Dental Surgery.

The Ninth Edition, with 320 *Illustrations.* Royal octavo.
Price, bound in cloth, bevelled boards, ... $6.00
" leather, ... 7.00

This edition of Dr. Harris's work has been subjected to a very thorough revision by competent professional gentlemen, and contains many and important additions, bringing the work fully up to the present state of dental science, and adding greatly to its value. The illustrations have also been much improved; some have been replaced by new drawings, and many new ones have been added. The publishers therefore offer it with the confident assurance that it will now be found a thorough elementary treatise, a text-book for the student, and a useful companion and guide for the practitioner.

Bond's Practical Treatise on Dental Medicine.

Being a Compendium of Medical Science, as Connected with the Study of Dental Surgery. By THOMAS E. BOND, M.D., *Professor of Special Pathology and Therapeutics in the Baltimore College of Dental Surgery.* The Third Edition. Octavo. $3.00

"We have spoken, or intended to speak, heartily in praise of Dr. Bond's work. It has unmistakable evidence of thorough medical science in its subject-matter, and of a capital authorship in its style and treatment."—*American Medical Journal.*

Robertson's Manual on Extracting Teeth.

Founded on the Anatomy of the Parts involved in the Operation, the Kinds and Proper Construction of the Instruments to be Used, the Accidents likely to occur from the Operation, and the Proper Remedies to be Used. By A. ROBERTSON, M.D., D.D.S., &c. A New Revised Edition. $1.50

"This work is valuable not only to the dental student and practitioner, but also to the medical student and surgeon."—*Dental Cosmos.*

Taft's Practical Treatise on Operative Dentistry.

A NEW EDITION, THOROUGHLY REVISED.

By JONATHAN TAFT, D.D.S., *Professor of Operative Dentistry in the Ohio College of Dental Surgery, &c. The Second Edition, thoroughly Revised, with additions, and fully brought up to the present state of the Science. Containing over* 100 *Illustrations.* Octavo. Leather, . . $4.50

"An examination of Mr. Taft's treatise enables us to speak most favorably of it. It is very thorough and very clear, showing that the author is practically familiar with the art which he teaches. The engravings are abundant and excellent, and, in fact, the whole mechanical execution of the volume is admirable, and reflects much credit on the publishers."—*Boston Medical and Surgical Journal.*

Fox on the Human Teeth.

Their Natural History and Structure, the Treatment of the Diseases to which they are Subject, the Mode of Inserting Artificial Teeth, &c. Edited by CHAPIN A. HARRIS, M.D., D.D.S., &c. *With* 250 *Illustrations.* Octavo. $4.00

Richardson's Practical Treatise on Mechanical Dentistry.

SECOND EDITION, MUCH ENLARGED.

By JOSEPH RICHARDSON, D.D.S., *Professor of Mechanical Dentistry in the Ohio College of Dental Surgery, &c. With over* 150 *beautifully executed Illustrations.* Octavo. Leather, ... $4.50

Handy's Text-Book of Anatomy,

AND GUIDE TO DISSECTIONS. *For the Use of Students of Medicine and Dental Surgery.* By WASHINGTON R. HANDY, M.D., *late Professor of Anatomy and Physiology in the Baltimore College of Dental Surgery. With* 312 *Illustrations.* Octavo. $4.00

Coles on Deformities of the Mouth.

Congenital and Acquired, with their Mechanical Treatment. By JAMES OAKLEY COLES, D.D.S., *Member of the Odontological Society, &c., &c.* Second Edition, Revised and Enlarged, with 8 Colored Engravings and 51 Illustrations on Wood. Price, ... $2 50

Heath on the Injuries and Diseases of the Jaws.

The Jacksonian Prize Essay of the Royal College of Surgeons of England, 1867. By CHRISTOPHER HEATH, F.R.C.S., *Assistant Surgeon to University College Hospital.* Over 150 Illustrations. Octavo. Price, ... $6 00

Tomes' System of Dental Surgery.

By JOHN TOMES, F.R.S., *Dentist to the Dental Hospital of London, Author of "Tomes' Dental Physiology," &c., &c. With* 208 *beautifully executed Illustrations.* Octavo. $4.50

Cooley's Toilet and Cosmetic Arts.

The Toilet and Cosmetic Arts, in Ancient and Modern Times. With a Review of the Different Theories of Beauty and copious allied Information, Social, Hygienic, and Medical, including Instructions and Cautions respecting the Selection and Use of Perfumes, Cosmetics, and other Toilet Articles; and a Comprehensive Collection of Formulæ, and Directions for their Preparation. By ARNOLD J. COOLEY, Author of "Cyclopædia of Receipts: Processes, Data, and Collateral Information, &c., in the Arts and Manufactures." With INDEX to about 5000 Matters of Interest, Use or Caution. Demi-Octavo. $3.00

Ott on the Manufacture of Soaps and Candles.

Including the Most Recent Discoveries, embracing all kinds of Ordinary Hard, Soft, and Toilet Soaps, especially those made by the Cold Process; and the Modes of Detecting Frauds in the Manufacturing and the Making of Tallow and Composite Candles. By ADOLPH OTT, Practical and Analytical Chemist. 12mo. With Illustrations. (Just ready.) $2.50

The author, in preparing this volume, has been careful to give a clear and concise account of the art of soap and candle making, as now practised, so as to make the work as practical in its character as possible. Appropriate illustrations have been added, and critical explanations of the various manipulations and mechanical arrangements, by which they are effected. Much new matter has also been incorporated in the book, never before published.

Piesse's Whole Art of Perfumery.

A NEW REVISED AND ENLARGED EDITION.

And the Methods of Obtaining the Odors of Plants; with Instructions for the Manufacture of Perfumes for the Handkerchief, Scented Powders, Odorous Vinegars, Dentifrices, Pomatums, Cosmetics, Perfumed Soaps, &c.; to which is added an Appendix, on Preparing Artificial Fruit Essences, &c. By G. W. SEPTIMUS PIESSE, Analytical Chemist. A new American from the Third London Edition. 12mo. With Numerous Illustrations. $3.00

DR. PIESSE's volume covers the entire ground of the subject upon which it treats. It is full of Useful and Curious Information, including also many Valuable Formulæ; and will be found of equal importance and interest to the practical man as to the general reader.

Overman's Practical Mineralogy, Assaying and Mining.

With a Description of the Useful Minerals, and Instructions for Assaying, according to the simplest Methods. By FREDERICK OVERMAN, Mining Engineer, &c. 12mo. $1.25

The object of this volume is to place before the public the characteristics and uses of minerals, in a popular style, avoiding, as far as possible, the use of scientific and technical terms. The subject is divided into three parts: — Mineralogy, or a Description of the Appearance of Minerals, with the localities in which they may or have been found; Assaying, or an Investigation of the value of Minerals, by means which are within the reach of every one; and Practical Mining in its simplest form.

Piggott on Copper Mining and Copper Ore.

Containing a full Description of some of the Principal Copper Mines of the United States, the Art of Mining, the Mode of Preparing the Ore for Market, &c., &c. By A. SNOWDEN PIGGOTT, M.D., Practical Chemist. 12mo. $1.50

Morfit's Chemical and Pharmaceutical Manipulations.

A Manual of the Chemical and Chemico-Mechanical Operations of the Laboratory. By CAMPBELL MORFIT, *Professor of Analytic and Applied Chemistry in the University of Maryland, assisted by* CLARENCE MORFIT, *Assistant Melter and Refiner in the United States Assay Office.* The Second Edition, Revised and Greatly Enlarged, with over 500 *Illustrations.* Octavo. $5.00

"The arrangement of the whole is such, that every student will be able to go through the work without a guide to lead him, provided the necessary apparatus are at his command. But even without them, a careful study of the book will give the attentive student a very useful insight in all the manipulations of the pharmaceutical chemist, and thousands, no doubt, who are prevented from attending the schools of pharmacy and chemistry, will gladly avail themselves of the only means left them for self-improvement. The amount of perseverance and industry displayed in the getting up of this work is truly astonishing, the clearness of expression in every sentence, and the accurateness of the 500 illustrations, are above praise. 'Morfit's Manipulations' ranges in utility immediately after the United States Dispensatory."—*Chemical Gazette.*

Branston's Hand-Book of Practical Receipts.

A Manual for the Chemist, Druggist, Medical Practitioner, &c., &c. Comprising the Officinal Medicines, their Uses, and Modes of Preparation, and Formulæ for Trade Preparations, Mineral Waters, Powders, Beverages, Dietetic Articles, Perfumery, &c.; with a Glossary of Medical and Chemical Terms, and a Copious Index. By THOMAS F. BRANSTON. *From the Second Revised and Enlarged Edition.* 12mo. $1.50

Campbell's Manual of Scientific and Practical Agriculture.

A Systematic Arrangement of all Scientific Knowledge bearing in any manner on the great work of Farming. For the use of Schools and Farmers. By PROF. J. L. CAMPBELL, *of Washington College, Va.* 12mo. *With Illustrations.* $1.50

This volume has been prepared to supply those already engaged in the culture of the soil with a guide, the study or perusal of which will enable them to improve upon the old system, or rather want of system, which has worn out so much of our best land, and has rendered the pursuit, in so many instances, unprofitable; and also to meet the demands of teachers for a text-book of the right kind, which will give the student such information as will fit him for the intelligent pursuit of agriculture as a business.

Darlington's Flora Cestrica;

OR, HERBORIZING COMPANION. *Containing all the Plants of the Middle States, their Linnæan Arrangement, a Glossary of Botanical Terms, a complete Index, etc.* By WILLIAM DARLINGTON, M.D. *The Third Edition, enlarged.* 12mo. $2.25

Miller on Alcohol, and Lizars on Tobacco.

Alcohol: Its Place and Power. By JAMES MILLER, F.R.S.E., *Professor of Surgery in the University of Edinburgh; President of the Medico-Chirurgical Society; Author of* MILLER'S *Principles and Practice of Surgery, etc., etc.* *The Use and Abuse of Tobacco.* By JOHN LIZARS, *late Professor of Surgery to the Royal College of Surgeons, etc., etc.* The Two Essays in One Volume. 12mo. $1.00

The first of these treatises was prepared by Prof. Miller at the request of the Scottish Temperance League, who were anxious to have a work of high authority, presenting the medical view of the Temperance question. It has passed through a great number of editions in Scotland, and has had a large sale in this country. The second was prepared by Prof. Lizars to show the pernicious consequences of excessive or habitual smoking. If purchased in *quantities*, either together or separately, by *Temperance* or other societies, they will be furnished at a reduced price

NEW SYDENHAM SOCIETY'S PUBLICATIONS.

LINDSAY & BLAKISTON, Philadelphia,

Are now prepared to receive subscriptions for the publications of THE NEW SYDENHAM SOCIETY for the year 1870, at Ten Dollars, payable in currency, and invariably in advance, and to furnish any of the previous years at the same rate and on the same terms.

The Practical Character and Permanent Value of these publications, and the very low price at which they are furnished, commend them to the favorable attention of the Medical Profession in the United States.

WORKS ALREADY PUBLISHED.

1859. (*First Year*.)

VOL. 1. DIDAY on Infantile Syphilis.
2. GOOCH on Diseases of Women.
3. MEMOIRS on Diphtheria.
4. VAN DER KOLK on the Spinal Cord, &c.
5. MONOGRAPHS (Kussmaul & Tenner, Græfe, Wagner, &c.)

1860. (*Second Year*.)

VOL. 6. Dr. BRIGHT on Abdominal Tumors.
7. FRERICHS on Diseases of the Liver. Vol. I.
8. A YEARBOOK for 1859.
9. ATLAS of Portraits of Skin Diseases. (1st Fasciculus.)

1861. (*Third Year*.)

VOL. 10. A YEARBOOK for 1860.
11. MONOGRAPHS (Czermak, Dusch, Radicke, &c.)
12. CASPER's Forensic Medicine. Vol. I.
14. ATLAS of Portraits of Skin Diseases. (2nd Fasciculus.)

1862. (*Fourth Year*.)

VOL. 13. FRERICHS on Diseases of the Liver. Vol. II.
15. A YEARBOOK for 1861.
16. CASPER's Forensic Medicine. Vol. II.
17. ATLAS of Portraits of Skin Diseases. (3d Fasciculus.)

1863. (*Fifth Year*.)

VOL. 18. KRAMER on Diseases of the Ear.
19. A YEARBOOK for 1862.
20. NEUBAUER and VOGEL on the Urine.

1864. (*Sixth Year*.)

VOL. 21. CASPER's Forensic Medicine. Vol. III.
22. DONDERS on the Accommodation and Refraction of the Eye.
23. A YEARBOOK for 1863.
24. ATLAS of Portraits of Skin Diseases. (4th Fasciculus.)

1865. (*Seventh Year*.)

VOL. 25. A YEARBOOK for 1864.
26. CASPER's Forensic Medicine. Vol. IV.
27. ATLAS of Portraits of Skin Diseases. (5th Fasciculus.)

1866. (*Eighth Year*.)

VOL. 28. BERNUTZ & GOUPIL on the Diseases of Women.
29. ATLAS of Portraits of Skin Diseases. (6th Fasciculus.)
30. HEBRA on Diseases of the Skin. Vol. I.
31. BERNUTZ & GOUPIL on Diseases of Women. Vol. II.

1867. (*Ninth Year*.)

VOL. 32. A BIENNIAL Retrospect of Medicine and Surgery.
33. GRIESINGER on Mental Pathology and Therapeutics.
34. ATLAS of Portraits of Skin Diseases. (7th Fasciculus.)
35. TROUSSEAU's Clinical Medicine. Vol. I.

1868. (*Tenth Year*.)

VOL. 36. THE COLLECTED WORKS OF Dr. ADDISON.
37. HEBRA on Skin Diseases. Vol. II.
38. LANCEREAUX's Treatise on Syphilis. Vol. I.
39. ATLAS of Portraits of Skin Diseases; (8th Fasciculus.)
40. A CATALOGUE of the PORTRAITS issued in the Society's Atlas of Skin Diseases. (Part I.)

1869. (*Eleventh Year*.)

VOL. 41. TROUSSEAU's Clinical Medicine. Translated and edited by Dr. Rose Cormack. Vol. II.
42. BIENNIAL RETROSPECT OF MEDICINE AND SURGERY, for 1867-8. Edited by Dr. Anstie, Dr. Barnes, Mr. Holmes, Mr. Power, Mr. Carter, and Dr. Underwood.
43. LANCEREAUX on Syphilis. Translated by Dr. Whitley. Vol. II., *completing the Work*.
44. A NINTH FASCICULUS of the ATLAS OF PORTRAITS OF SKIN DISEASES

WORKS TO BE PUBLISHED IN 1870.

Trousseau's Clinical Medicine. Vol. III.
Stricker's Manual of Histology. Vol. I.
Niemeyer on Phthisis.
Wunderlich's Treatise on the Use of the Thermometer in Disease.
A Tenth Fasciculus of the Atlas of Skin Diseases.

Subscribers at a distance can have their Volumes mailed to them, postage paid, as they appear, by remitting $1 50 in addition to the subscription price for the year.

Non-Subscribers can obtain the books published during any one year by subscribing and paying for that year $10.00, but no volumes or books can be had otherwise or separately except the following:

The Year Books for 1859, '60, '61, and '62, for $10 50
Portraits of Skin Diseases. Fasciculi 1 to 9, for 42 00

A Descriptive Catalogue of the Society's Atlas of Portraits of Diseases of the Skin, and their last REPORT, will be furnished gratis upon application.

AMERICAN & BRITISH PERIODICALS
SUPPLIED BY
LINDSAY & BLAKISTON,
PHILADELPHIA.
Subscriptions Payable in Advance.

NAMES.	WHERE PUBLISHED.	PRICE Per Annum.
YEARLY.		
The Pennsylvania Hospital Reports,	Vols. 1 & 2, each,	$4 00
The Physician's Visiting List, for 1871, prices reduced.	See Catalogue.	
Clinical Society's Transactions,	London,	
St. Andrew's Medical Graduates' Association Reports,		
The London Hospital Reports,	London,	
The Liverpool " "	"	
St. George's " "	"	
St. Bartholomew " "	"	
Guy's " "	"	
Obstetrical Society's Transactions,	"	
Pathological " "	"	
Medico-Chirurgical Society's Transactions,	"	
The New Sydenham Society's Publications. 3 to 4 volumes published annually,	"	10 00
HALF-YEARLY.		
Braithwaite's Retrospect of Medicine and Surgery,	Reprint,	2 50
Ranking's Half-Yearly Abstract " "	"	2 50
QUARTERLY.		
American Journal of the Medical Sciences,	Philadelphia,	5 00
British and Foreign Medico-Chirurgical Review,	London,	10 00
The Dublin Quarterly Journal of Medicine,	Dublin,	10 00
American Journal of Syphilography and Dermatology,	New York,	3 00
Journal of Anatomy and Physiology. $2.00 per Number,	London,	
Microscopical Journal,	"	8 00
The American Journal of Obstetrics,	New York,	4 00
Journal of Psychological Medicine,	"	5 00
The New Orleans Journal of Medicine,	New Orleans,	5 00
The Ophthalmic Hospital Reports,	London,	
BI-MONTHLY.		
The American Journal of Pharmacy,	Philadelphia,	3 00
MONTHLY.		
Journal of the Gynæcological Society,	Boston	3 00
The London Lancet,	Reprint,	5 00
The Pharmaceutical Journal,	London,	6 00
The Medical and Surgical Journal,	Edinburgh,	8 00
The Practitioner. Edited by F. E. Anstie, M. D.,	Reprint,	4 00
Chemical News,	New York,	5 00
The Medical Archives,	St. Louis,	3 00
The Richmond and Louisville Medical Journal,	Louisville,	5 00
The Chicago Medical Journal,	"	3 00
New York Medical Journal,	New York,	4 00
SEMI-MONTHLY.		
The Medical Times,	Philadelphia,	4 00
The Medical Record,	New York,	4 00
WEEKLY.		
The London Lancet,	London,	12 00
Medical Times and Gazette,	"	12 00
British Medical Journal,	"	12 00
Medical and Surgical Reporter,	Philadelphia,	5 00
Medical and Surgical Journal,	Boston,	4 00

Any other Journals will be furnished to order.

English Medical Books. Recent Importations.
MANY OF THEM ON HAND IN QUANTITIES.
Sent free by mail at the prices annexed, and furnished to the Trade at a liberal discount.

SOELBERG WELLS' Complete Treatise on Diseases of the Eye. Second LONDON Edition,	$6.50
WELLS' on Long, Short, and Weak Sight. 3d Edition,	3.00
HEATH on Diseases and Injuries of the Jaws. Illustrated,	6.00
HOLMES' Surgical Diseases of Children. 2d Edition,	9.00
GREENHOW on Chronic Bronchitis,	2.00
BEALES' How to Work the Microscope. 400 Illustrations,	7.50
BEALE on the Microscope in Practical Medicine. 500 Illustrations,	7.00
BEALE on Kidney Diseases, Urinary Deposits, &c., &c. 400 Illustrations,	10.00
COLES on Deformities of the Mouth. Second Edition. Colored Illustrations,	2.50
MEDICINE IN MODERN TIMES. 8vo.	2.50
TROUSSEAU'S Lectures on Clinical Medicine. Vols. 1, 2, and 3, *each*,	5.00
THE NEW SYDENHAM SOCIETY'S PUBLICATIONS, per annum,	10.00
TAYLOR's Principles and Practice of Medical Jurisprudence,	14.00
KIRKES' Hand-Book of Physiology. 7th London Edition,	5.00
BEALE on Diseased Germs. 50 Colored Engravings,	
LEGG's Guide to the Examination of Urine. 2d Edition,	75
MEDICO CHIRURGICAL SOCIETY'S TRANSACTIONS. Vol. 52, for 1869.	6.00
ALTHAUS' Medical Electricity. 2d Edition. Enlarged and Illustrated,	5.00
BASHAM on Dropsy. 3d Edition. Enlarged and Illustrated,	6.25
COLLIS on Cancer and Tumors. Illustrated,	7.00
COOKE on Cancer; its Allies, &c. Colored Illustrations,	6.25
PARKE's Practical Hygiene. 2d Edition,	8.00
THUDICUM's Pathology of the Urine. Illustrated,	7.00
GRAVE's Clinical Lectures by Neligan. A New Edition,	6.00
DAY's Clinical Histories with Comments.	3.75
BEALE's Protoplasm; or, Life, Matter, Mind. 2d Edition. 8 Illustrations,	3.00
TOYNBEE on the Ear. A new Edition, by Hinton.	5.00
SALT's Mechanical Treatment of Deformities. Illustrated,	7.50
HOLMES' System of Surgery, 5 vols., 8vo. New Edition, per vol.,	9.00
WEBER's Hand-Book of Auscultation and Percussion,	1.00
LEE's Diseases of the Veins and Hemorrhoidal Tumors,	4.00
CATLOW's Principles of Æsthetic Medicine,	4.50
SOUTHWOOD SMITH's Philosophy of Health. 11th Edition,	3.75
BAKER BROWN on Ovarian Dropsy. 8vo., cloth,	2.50
HARLEY's Old Vegetable Narcotics. Octavo,	4.00
HIGGINBOTTOM on the Use of Nitrate of Silver,	3.00
WAULTUCH's Dictionary of Materia Medica and Therapeutics. Octavo,	7.50
REYNOLD's System of Medicine. New Edition, per volume,	6.00
HILTON's Lectures on Rest and Pain. Illustrated,	8.00
FOWNE's Manual of Chemistry. 10th Edition. Revised and Enlarged,	
LITTLE on Spinal Weakness and Curvatures,	2.50
HOBLYN's Dictionary of Medical Terms. 9th Edition. Enlarged,	6.25
SANDERSON's Hand Book of the Sphygmograph,	1.75
MORRIS on Irritability, &c. 18mo., cloth,	2.25
BIRCH on the Use an Value of Oxygen. 16mo., cloth,	1.75
WRIGHT on Uterine Disorders. 8vo., cloth,	3.75
GANT on the Irritable Bladder: its Causes and Curative Treatment,	2.50
URQUHART's Manual of the Turkish Bath. Illustrated,	2.50
GUY's Forensic Medicine. 3d Edition. Revised and Enlarged, with Illustrations,	6.25
GAIRDINER on Gout: its History, Causes, and Cure,	4.25
MOORE on Rodent Cancer,	3.00
GRIFFIN's Chemical Recreations. Illustrated,	6.25
SANKEY's Lectures on Mental Diseases. 8vo., cloth,	4.00
SYNOPSIS of the Pathological Series in the Oxford Museum,	3.75
HUMPHRY on the Human Skeleton. Octavo. Illustrated,	8.00
TAYLOR on Poisons. 2d London Edition,	6.25
ROYLE & HEADLAND's Materia Medica and Therapeutics. 5th Edition,	6.25
LONDON Obstetrical Society's Transactions. Vol. 10,	7.50
POWERS on the Eye. Illustrated. Octavo, cloth,	10.00
CARPENTER on the Microscope. Fourth Revised Edition,	6.25

See Periodical List of **Hospital Reports, Transactions of Societies, and Foreign Medical Periodicals,** furnished at the prices annexed.

☞ **English Medical Books** not on hand imported promptly to order at the lowest rates.

LINDSAY AND BLAKISTON'S PUBLICATIONS.

Anatomical Plates.

MARSHALL'S PHYSIOLOGICAL DIAGRAMS. Nine Plates in the set. Figures six feet long; Life-size; printed on one sheet 4x7 feet; beautifully colored.

Contents.

No. 1.—The Skeleton and Ligaments.
2.—The Muscles, Joints, and Animal Mechanics.
3.—The Viscera in Position; the Structure of the Lungs.
4.—The Heart and Blood-vessels.

No. 5.—The Lymphatics, or Absorbents.
6.—The Digestive Organs.
7.—The Brain and Nerves.
8.—The Organs of the Senses.
9.—The Textures and Microscopic Structure.

Handsomely mounted on rollers. Price, ... $90.00
In sheets, ... 60.00

LECTURE DIAGRAMS FOR INSTRUCTION IN PREGNANCY AND MIDWIFERY. 20 Plates of the largest imperial size, printed in colors. Drawn and Edited with Explanatory Notes by Dr. B. S. Schultze, Professor of Midwifery at the University in Jena. Price, . . . $18.00

BOURGERY & JACOB'S PLATES. Twenty in the set. Figures three feet long, beautifully colored; mounted on rollers.
Per set. Price, ... $50.00
Sold Separate, each, ... 3.00

The following are the subjects and arrangements of the plates:—

Osteology and Syndesmology.

Plate I. Anterior Plane. *Right side:* The Dry Bones. *Left side:* The Bones clothed with their ligaments.
Plate II. Posterior plane. The same arrangement.

Myology and Aponeurology.

Plate III. Anterior plane. *Right side:* Superficial muscles. *Left side:* Superficial aponeuroses.
Plate IV. Anterior plane. *Right side:* Muscles of the second layer. *Left side:* Muscles of the third layer.
Plate V. Posterior plane. *Right side:* Superficial muscles. *Left side:* Superficial aponeuroses.
Plate VI. Posterior plane. Second and Third layer of muscles.
Plate VII. Lateral plane. Superficial and deep muscles. Muscles of the os hyoides.
Plate VIII. Diaphragm. Interior of the trunk, muscles of the lower jaw, of the tongue, of the vlum palati, and of the pharynx.

Angiology.

Heart, lungs, arteries, veins, and lymphatics. On the different figures are indicated the points at which compression on the ligature of the vessels is effected, and in regard to the veins in particular, the proper points for performing venesection.

Plate IX. Interior of the trunk. Heart, lungs, and their envelopes. Large vessels.
Plate X. Vessels of the thorax and abdomen, azygos vessels, cerebral and spinal venous sinuses.
Plate XI. Anterior plane. Sub-cutaneous vein, and deep vessels.
Plate XII. Posterior plane. Superficial veins, and deep vessels.
Plate XIII. Lateral plane. Partial figures, internal maxillary and internal carotid vessels, &c.
Plate XIV. Lymphatic vessels.

Neurology.

Plate XV. Anterior plane. Encephalic nerves. Nerves of the extremities.
Plate XVI. Posterior plane. Studies of the ganglions and their nerves. Studies of the fifth and seventh cerebral pairs.
Plate XVII. Brain, spinal marrow, and envelopes. Organs of the senses. Larynx.

Digestive Apparatus.

Plate XVIII. Alimentary canal; stomach, intestines, chylliferous vessels, peritoneum.
Plate XIX. Stomach, liver, pancreas, spleen, kidneys, supra-renal capsules, bladder. Abdominal venous system. Great sympathetic and pneumo-gastric nerves.
Plate XX. Complete study of the perineum in both sexes. Male and female organs of reproduction. Embryotomy.

FIEDLER'S ANATOMICAL PLATES. 4 in the set, mounted on rollers. Price, . . $15.00

LAMBERT'S ANATOMICAL PLATES. Figures three feet long, handsomely colored, mounted on rollers. Price, ... $15.00

LINDSAY & BLAKISTON'S PUBLICATIONS.

For Sale by Booksellers in the principal Cities of the Union, or sent by mail, free of postage, upon receipt of the retail price.

Aitken's Science and Practice of Medicine, from the fifth London edition, with American additions equal to 500 pages of the Lond. edit., 2 vols. 8vo. $12.00
Althaus's Medical Electricity. New Edition 5.00
Acton on the Reproductive Organs . 3.00
Anstie on Stimulants and Narcotics . 3.00
Byford on Diseases of Women . . 5.00
Biddle's Materia Medica. Third edition. 4.00
Branston's Practical Receipts . . . 1.50
Beale's How to Work the Microscope . 7.50
" Microscope in Practical Medicine 7.00
Beale on the Kidneys, Urine, &c. 400 Ill. 10.00
Beasley's Book of Prescriptions . . 4.00
" Druggist's Receipt Book. . 3.50
Barth & Roger's Auscultation . . . 1.25
Birch on Constipation. Third edition 1.00
Bouchardat's Annual of Therapeutics,&c. 1.50
Bull's Maternal Management of Children 1.25
Braithwaite's Epitome, 2 vols. . . . 10.00
Beale's Protoplasm, 2d edition, plates, 3.00
" On Disease Germs, do. 1.75
Hodge on Fœticide. Paper 30 cts.; cloth, .50
Chambers' Lectures. Renewal of Life 5.00
Chew on Medical Education . . . 1.00
Cohen's Therapeutics of Inhalation . 2.50
Cazeaux's Obstetrical Text-Book, the Fifth American edition, very much enlarged. 175 Illustrations . . . 6.50
Cleaveland's Pronouncing Medical Lexicon. Eleventh edition 1.25
Carson's History Medical Department University of Pennsylvania . . . 2.00
Goff's Physician's Day-Book, Ledger, &c. 12.00
Dixon on Diseases of the Eye, new ed. 2.50
Kirkes Manual of Physiology, 7th Lon.ed. 5.00
Legg's Guide to the Examination of Urine, .75
Durkee on Gonorrhœa and Syphilis, Fifth edition, revised and improved . . 5.00
Fuller on Rheumatism. A new edition.
Garratt on Medical Batteries . . . 2.00
Graves' Clinical Medicine. New ed. . 6.00
Greenhow on Chronic Bronchitis . . 2.00
Gross' American Medical Biography . 3.50
Headland on the Action of Medicine . 3.00
Heath's Diseases and Injuries of the Jaws 6.00
Hewitt on the Diseases of Women . . 5.00
Hilles' Pocket Anatomist 1.00
Holmes' Surgical Diseases of Children 9.00
Hufeland's Art of Prolonging Life . . 1.25
Hillier's Diseases of Children . . . 3.00
Mackenzie on the Laryngoscope, Rhinoscopy, and Diseases of the Throat . 3.00
Morris on Scarlet Fever 1.50
Meigs' & Pepper's Treatise on Diseases of Children. Fourth edition, rewritten and very much enlarged 6.00
Maxson's Practice of Medicine . . . 4.00
Mendenhall's Medical Student's Vade Mecum. Eighth edition 2.50
Pennsylvania Hospital Reports. Vols. 1 and 2, each 4.00
Paget's Lectures on Surgical Pathology 6.00
Pereira's Physician's Prescription Book 1.25
Physician's Visiting List. Various sizes and prices. See Catalogue.

Prince's Orthopedic Surgery . . . $3.00
Prince's Plastic Surgery. Illustrated 1.50
Renouard's History of Medicine . . 4.00
Radcliffe on Epilepsy, Pain, Paralysis,&c. 2.00
Lawson's Complete Text-Book of Diseases and Injuries of the Eye . . 2.50
Ruppaner on Laryngoscopy, &c. . . 2.00
Ryan's Philosophy of Marriage . . 1.00
Reese's Analysis of Physiology . . . 1.50
Reese's American Medical Formulary 1.50
Sydenham Society's Biennial Retrospect 2.00
Stillé's Epidemic Meningitis . . . 2.00
Sansom on Chloroform, its Action, Modes of Administration, &c., &c. . . . 2.00
Stokes on Diseases of the Heart . . 3.00
Spratt's Obstetric Tables. 4to; col'd Pl. 8.00
Skoda on Auscultation and Percussion 1.50
Sydenham Society's Pub. Per year, 10.00
Tyson's Cell Doctrine. Illustrated. . 2.00
Tanner's Practice of Medicine, 5th ed. 6.00
Tanner on Diseases of Children . . 8.00
Tanner's Index of Diseases 3.00
Tanner's Memoranda of Poisons . . .50
Trousseau's Clinical Medicine. Vols. 1, 2, and 3, each 5.00
Thompson on Pulmonary Consumption 1.25
Tilt's Elements of Female Hygiene . 1.50
Taylor's Movement Cure 1.50
Virchow's Cellular Pathology . . . 5.00
Sœlberg Wells on the Eye. 2d London Edition, with Illustrations . . . 6.50
Walker on Intermarriage 1.50
Wythe's Pocket Dose and Symptom Book. Eighth edition 1.00
Waring's Practical Therapeutics . . 6.00
Walton's Operative Ophthalmic Surgery 4.00
Watson's Practice, Abridged . . . 2.00
Wright on Head-Aches 1.25
Wells on Long, Short, and Weak Sight. Third edition 3.00
Weber's Clinical Hand-Book of Auscultation and Percussion 1.00
Harris' Dictionary of Medical Terminology and Dental Surgery . . . 6.50
Harris' Principles and Practice of Dental Surgery. Ninth edition . 6.00
Bond's Dental Medicine 3.00
Robertson on Extracting Teeth . . . 1.50
Taft's Operative Dentistry 4.50
Fox on the Human Teeth 4.00
Richardson's Mechanical Dentistry . 4.50
Handy's Text-Book of Anatomy . . 4.00
Coles on Deformities of the Mouth. 2d Edition. Colored Illustrations . 2.50
Tomes' System of Dental Surgery . 4.50

SCIENTIFIC.

Cooley's Toilet and Cosmetic Arts . . 8.00
Ott on the Manufacture of Soap and Candles 2.50
Piésse on Perfumery. A new edition 3.00
Overman's Mineralogy, Assaying, and Mining 1.25
Piggott on Copper Mining 1.50
Morfit's Chemical Manipulations . . 5.00
Campbell's Agriculture 1.50
Darlington's Flora Cestrica 2.25
Miller & Lizars on Alcohol and Tobacco 1.00

CLASSIFIED AND DESCRIPTIVE CATALOGUES OF MEDICAL BOOKS, WITH THE PRICES ANNEXED, SENT FREE BY MAIL UPON APPLICATION.

www.ingramcontent.com/pod-product-compliance
Lightning Source LLC
Chambersburg PA
CBHW032121230426
43672CB00009B/1819